Essential Animals

Justin Gerlach

ISBN: 978-1-716-94879-4

Contents

1.	Introduction and historical phylogeny	5
2.	Three clades and ancestry	7
3.	Porifera (sponges)	9
4.	Placozoa	11
5.	Ctenophora (comb-jellies)	13
6.	Cnidaria (sea anemones, jellyfish & myxozoans)	15
7.	Xenacoelomorpha (basal flatworms)	17
8.	Orthonectida	19
9.	Dicyemida	21
10.	Chaetognatha (arrow worms)	23
11.	Gnathostomulida (spiny jaws)	25
12.	Rotifera (wheel animals + spiny-headed worms)	27
13.	Micrognathozoa	29
14.	Gastrotricha (hairybellies)	31
15.	Platyhelminthes (flatworms)	33
16.	Cycliophora	35
17.	Annelida (segmented worms)	37
18.	Mollusca	39
19.	Nemertea (ribbon worms)	41
20.	Ectoprocta + Entoprocta (moss and goblet animals)	43
21.	Brachiopoda (lamp shells)	45
22.	Phoronida (horseshoe worms)	47
23.	Kinorhyncha (mud dragons)	49
24.	Priapulida (penis worms)	51
25.	Loricifera (jewel animals)	53
26.	Nematoda (round worms)	55
27.	Nematomorpha (Gordian worms)	57
28.	Onychophora (velvet worms)	59
29.	Tardigrada (water bears)	61
30.	Arthropoda (arthropods)	63
31.	Deuterostomia – embryology and groups	65
32.	Echinodermata (echinoderms)	67
33.	Hemichordata (acorn worms & sea angels)	69
34.	Chordata (lancelets, sea squirts and vertebrates)	71
35.	References	73

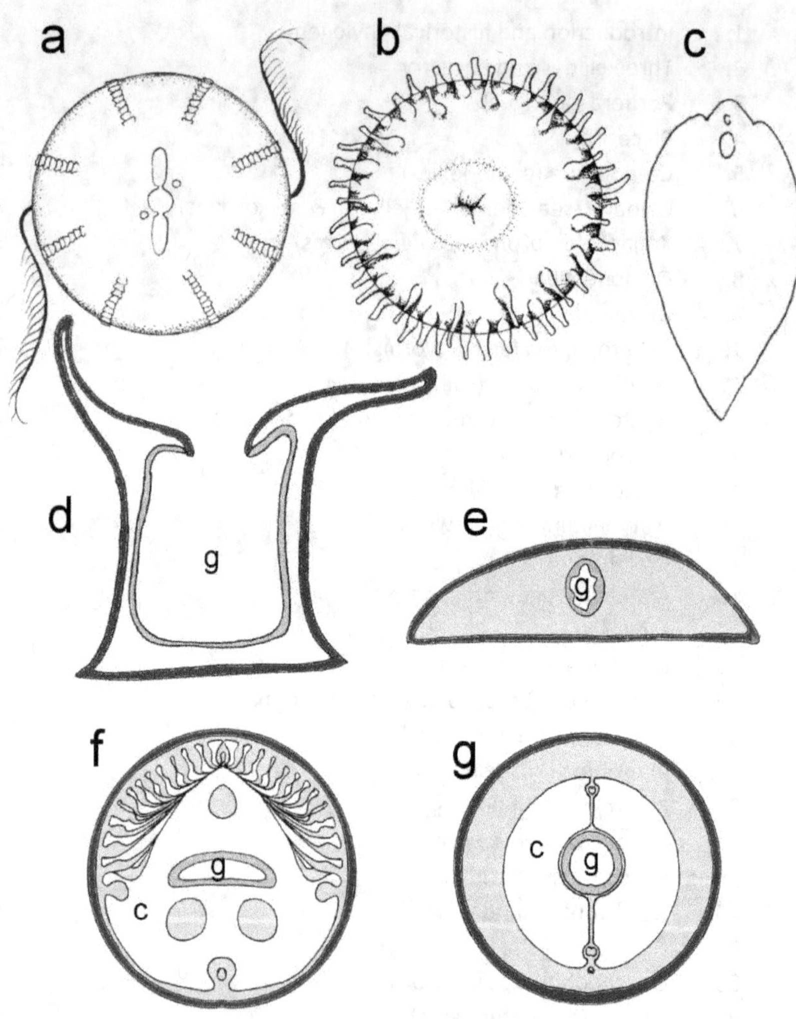

Basic features of animal structure: symmetry (a-c), tissue layers (d-e), body cavities (e-g): a) rotational, b) radial, c) bilateral, d) diploblast, e) triploblast acoelomate, f) pseudocoelomate, g) coelomate. Tissues shaded: dark grey – ectoderm, mid grey – endoderm, light grey - mesoderm; c = coelom, g = gut. Phyla: a) Ctenophora, b & d) Cnidaria, c & e) Platyhelminthes, f) Nematoda, g) Annelida

1. Introduction and historical phylogeny

This book aims to provide what it says: the essential information on the Animal Kingdom. Very few university biology courses cover animal diversity and phylogenetics in any comprehensive way. Few, if any, mention all animal phyla, and only in a very cursory manner, concentrating on the most familiar arthropods and chordates. This book covers all the animal phyla, with a brief digest of their ecology, main morphological features and evolutionary history. Other books, especially Giribert & Edgecombe (2020) are recommended for more detail.

Morphology and the evolutionary relationships are important for understanding animal diversity. Fundamental morphological features comprise symmetry, tissue layers, development of the digestive system and the body cavity. This links to the evolutionary relationships through developmental genetics, especially the Hox genes (Chapter 4).

Most animal groups are bilaterally symmetrical with distinct front and back ends, but two phyla are radial, one rotational and two are asymmetrical. There is a rough correlation between Hox gene arrangement and symmetry, and with structural complexity.

Most animal groups have distinct tissue layers which evolve sequentially. The simplest are diploblasts (two-layered) with an external ectoderm and an internal endoderm forming the gut. More derived groups have a third layer, the mesoderm, between them. The endoderm digestive system originates as a blind-ended, or 'bottle', gut but in most phyla it develops into a conventional through-gut with a separate mouth and anus.

Most animals have a body cavity, an open, fluid-filled space that is totally enclosed within the body. This may be present as a fully open cavity, largely filled in with organs, or modified into a circulatory system. Animals have been divided into three categories on the basis of this feature: acoelomates lacking such a cavity, pseudocoelomates with a cavity between the endoderm and mesoderm, and coelomates with the cavity surrounded by mesoderm. It develops either as a split within the body, or as a pocket budding off from the endoderm. The development seems to be largely convergent (see Chapters 8 and 30).

The arrangement of the animal phyla covered here is based on the most recent published phylogenies. Although this has some points of uncertainty, the phyla themselves are relatively well established. Their defining characters ('synapomorphies': evolutionary innovations shared by members of the group) are summarised here.

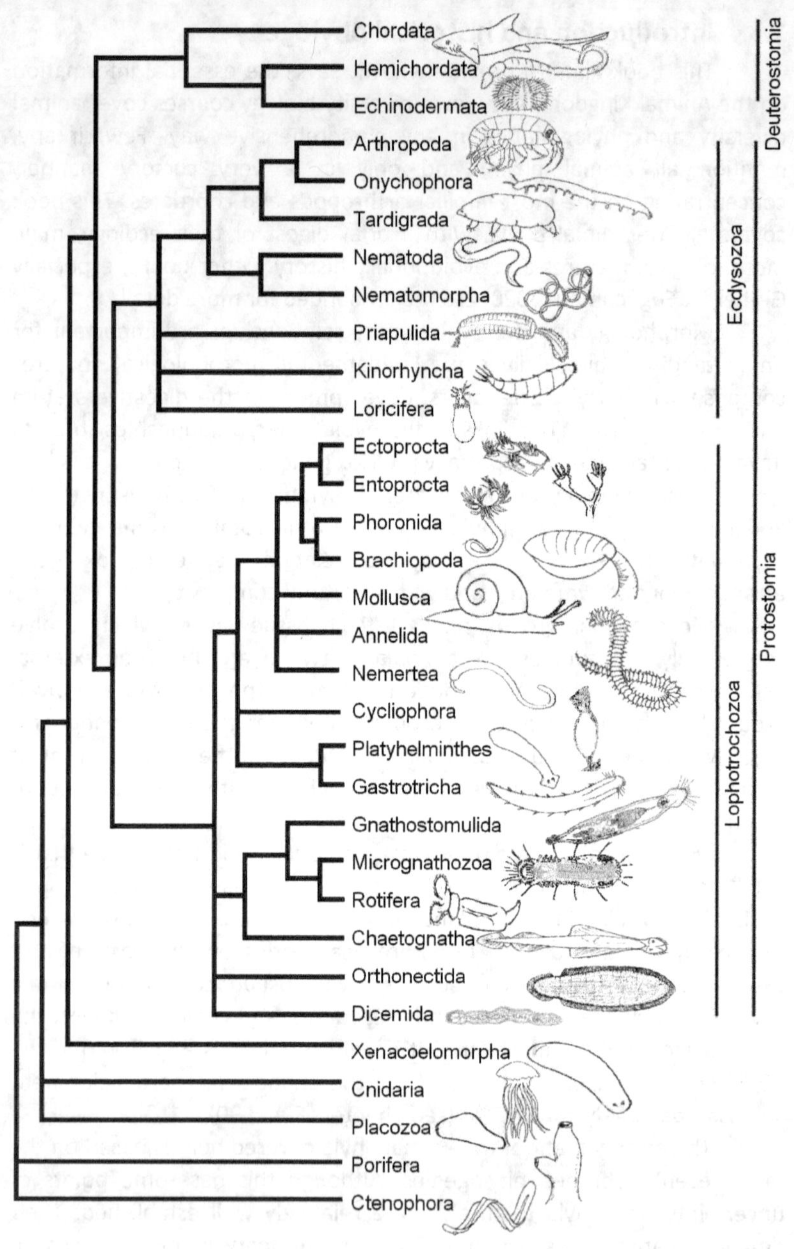

Chordata
Hemichordata
Echinodermata
Arthropoda
Onychophora
Tardigrada
Nematoda
Nematomorpha
Priapulida
Kinorhyncha
Loricifera
Ectoprocta
Entoprocta
Phoronida
Brachiopoda
Mollusca
Annelida
Nemertea
Cycliophora
Platyhelminthes
Gastrotricha
Gnathostomulida
Micrognathozoa
Rotifera
Chaetognatha
Orthonectida
Dicemida
Xenacoelomorpha
Cnidaria
Placozoa
Porifera
Ctenophora
Deuterostomia
Ecdysozoa
Protostomia
Lophotrochozoa

2. Three clades and ancestry

Animals are a well-defined group, recognised as heterotrophic multi-cellular eukaryotes with a well-developed pattern of embryonic development. They are unique in possessing collagen, characteristic sperm structure and highly reduced mitochondrial genomes.

Our concept of the Animalia has changed over the past 150 years. In 1894 Animalia was proposed to comprise Protozoa, Metazoa and Porifera. The single-celled protozoans are now excluded from the animals, as for a while were the sponges (Porifera). Excluding sponges is no longer accepted and now Metazoa is identical to the Animalia.

Historically, animal relationships were viewed as a pattern of increasing complexity, either explicitly or implicitly, leading to the perfection of the vertebrates. In this scheme the simplest groups would be the most basal and complexity could be seen as the progressive addition of morphological features. This is now known to be incorrect; all living organisms are regarded as being equally evolved and any implication of 'higher' and 'lower' groups should be avoided as much as possible. The use of terms such as 'basal' or 'derived' are convenient short-hand for the groups that are at the base of the tree in the particular way it is drawn. That is not a fixed position however and any particular topology is chosen simply for practical (or even aesthetic) reasons.

Since 1997 molecular phylogenies have dramatically changed our view of animal phylogeny, splitting our traditional scheme into the 'three-clades' model of Deuterostomia, Lophotrochozoa and Ecdysozoa. A compilation of the most recent versions is used here (shown opposite).

Our current view of animal evolution places us within the opisthokont eukaryotes, along with various protists and fungi. The sister-group of animals is the Choanoflagellata, solitary or colonial aquatic protists. These resemble sponge choanocyte cells (Chapter 3) and are similar to animals in having the genes for cell adhesion and signalling. Animals have many more transcription factors, including the Hox genes.

Animal fossils go back to the Ediacaran (555 million years ago), but molecular clock estimates indicate an origin about 750 million years ago.

References: Aguinaldo et al. 1997; Giribert & Edgecombe 2020; Lozano-Fernandez et al. 2017

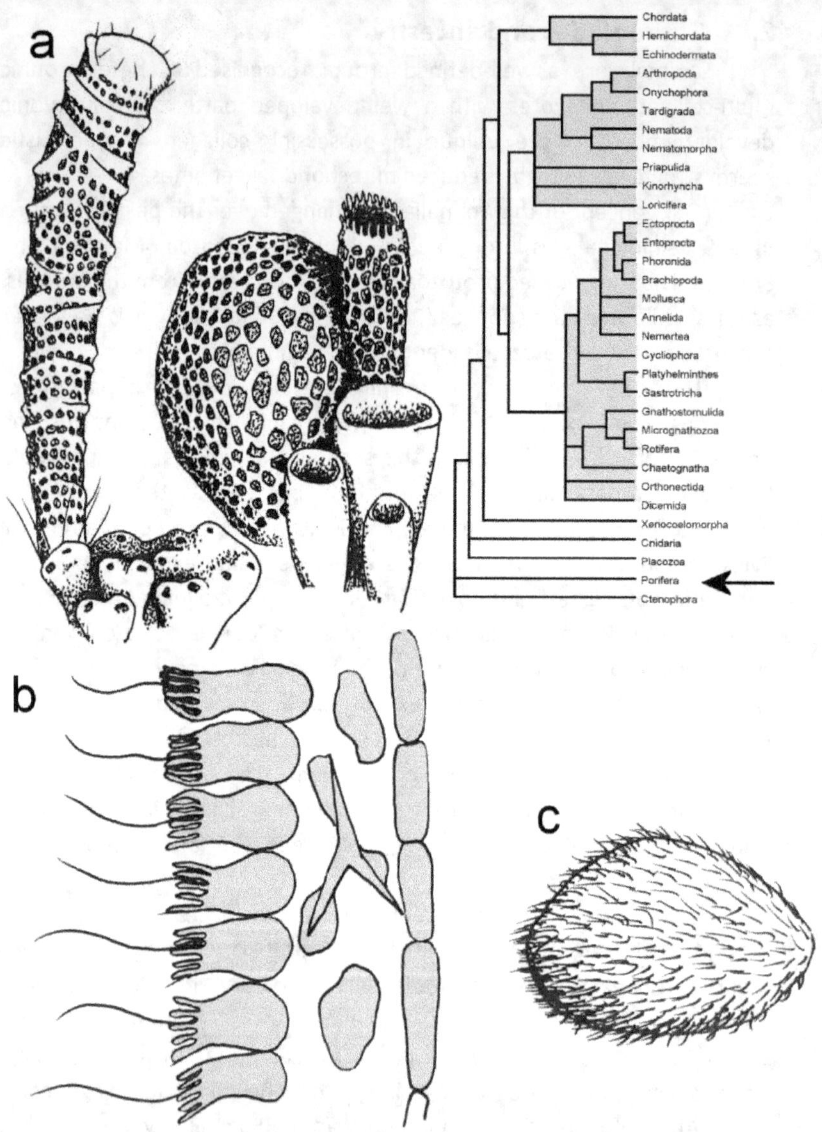

Porifera: a) range of morphologies, b) section through the body wall, left – choanocytes, centre – amoebocytes and spicules, right – pinacocytes, c) larva

3.   **Porifera** (sponges)

Porifera are named for their structure: porus being Latin for pore, although the Latin word for a sponge is used as a genus: *Spongia*.

Sponges are almost all marine (8,300 species), but a small number of species (50) live in freshwater. They are filter-feeders from the subtidal zone to the deep sea.

Structurally sponges can be regarded as the simplest of animals, lacking permanent tissue development. They have three main cell types that form distinct 'tissue' layers: external protective pinacocytes, internal choanocyte collar-cells and interstitial archaeocytes. The choanocytes actively pump water through the sponge using flagellae. These layers can redevelop if the cells are dissociated, meaning that the sponge's structure is only temporary, and develops depending on the external environment. Thus all sponge cells are effectively stem-cells throughout life, and sponges could be regarded as intermediate between coloniality and true multicellularity.

The tissues are supported by collagen fibres or by skeletal silica or calcium carbonate spicules. A small number of species are able to use the spicules as hook to trap prey, and are categorised as carnivorous. Other than these carnivorous sponges, the sponge structure typically takes the form of a cup or flask, with an open central atrium. Water flows through pores into the atrium and is expelled through an apical 'osculum'.

Reproduction may be asexual by budding or fragmentation, or sexual. Gametes are released into the water column and fertilisation results in a planktonic larva (with a few species having direct development). Larvae are varied in form.

Sponges were long assumed to be the most basal of animal phyla, although this is now uncertain due to recent molecular studies suggesting that ctenophores are more basal (Chapter 5). Although sponges are structurally the simplest animals it has been found that their genomes contain genes for relatively complex structures such as neurones. It is not impossible that some species may use these genes; traditionally it has been assumed that sponges with their limited tissue structure lacked an epithelium but it is now known that some sponges possess the type IV collagen that is found in the basement membranes of epithelia.

Fossil sponges date back to the early Cambrian 540 million years ago, possibly the Ediacaran 600 million years ago.

References: Boute et al. 1996

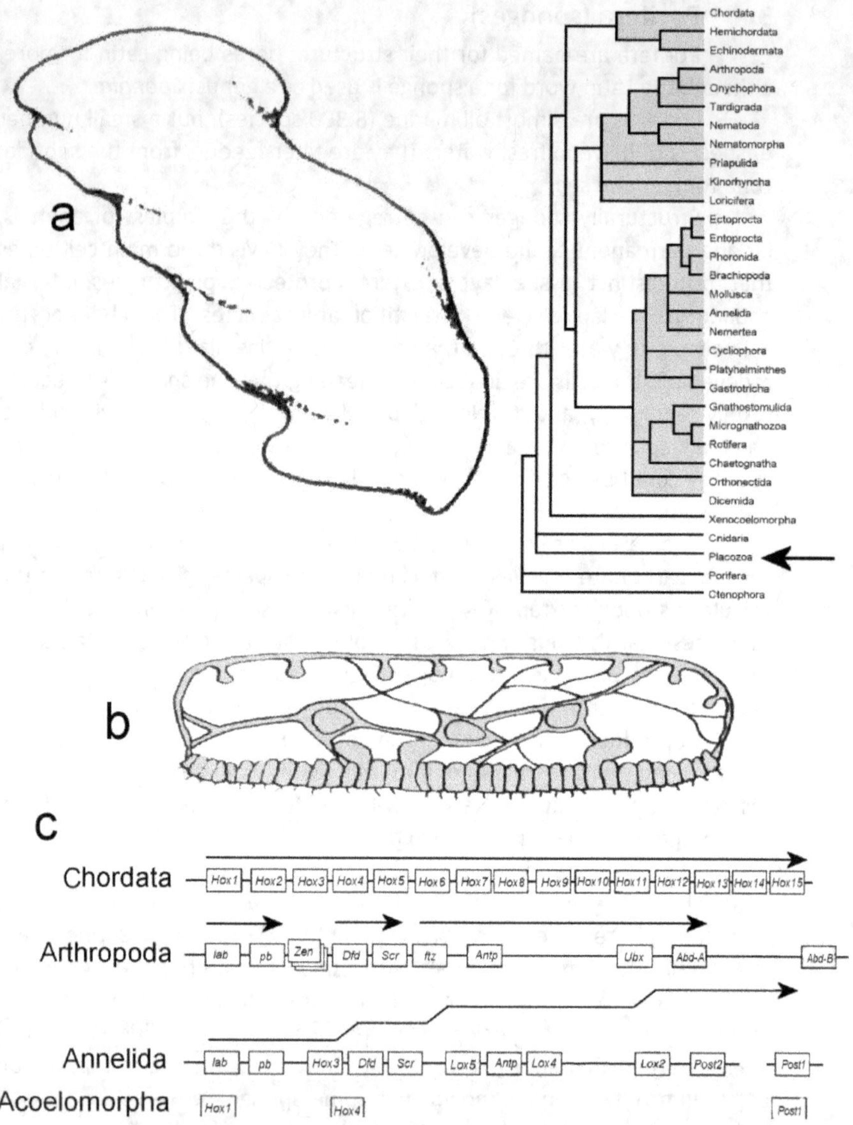

Placozoa: a) morphology, b) section, c) Hox gene expression in different animals – colinear, simultaneously colinear, simultaneously scattered. Arrows show the sequence of expression

4.   Placozoa

Three species of Placozoa have been described, but genetic studies have identified 19 separate genetic forms. All are small animals (under 3 mm), simple and disc-shaped. They are marine benthic animals found in different habitats in all oceans. Very little is known of their biology.

Structurally they have no distinct symmetry and comprise just two cell layers, with six distinct cell types. Cells on the lower surface are ciliated, those on the upper are flattened. Between these are occasional interconnected fibre cells and secretory cells. Although they are capable of coordinated movement there is no nervous system (but neuropeptide secreting cells are present). They feed on algae by extracellular digestions; there is no gut. Reproduction is by fragmentation and budding. There is no evidence of sexual reproduction.

They have always been of interest to animal phylogenetics; the name Placozoa is derived from the hypothetical ancestral 'Urmetazoan' animal, named a 'Placula'. Placozoans have been suggested to be closer to more complex animals than to sponges due to the presence of a ParaHox gene, forming the 'ParaHoxozoa' along with cnidarians and the bilateral animals. However a ParaHox gene may be present in some sponges.

ParaHox genes and the better known Hox may have arisen from duplication of an ancestral 'ProtoHox' gene. These homeobox transcription factors interact with structural genes to control the timing and location of organ development. An odd features of Hox genes is that they are expressed along the body in the same order that they are found on chromosomes. The significance of this colinearity is obscure, and it is not always the case; some groups have shuffled these genes and, perhaps most extremely, in octopus each Hox gene is on a separate chromosome. Hox colinearity is often lost, or is incomplete. In fact, full colinearity is only known from Hemichordata, Chordata (except Urochordata), Mollusca and Arthropoda. Even in the last two there are exceptions.

Despite extreme simplicity placozoans do have genes for the extracellular matrix of epithelia (including collagen IV). Similarly, although microRNAs are absent from placozoans, they have the genes for processing microRNAs (small molecules that regulate both translation and degradation of messenger RNAs). These observations suggest that at least some features of placozoans are secondarily simplified.

References: Albertin et al. 2015; Eitel et al. 2013; Holland 2013; Pastrama et al. 2019; Ryan et al 2010; Schiemann et al. 2017; Srivastava et al. 2008

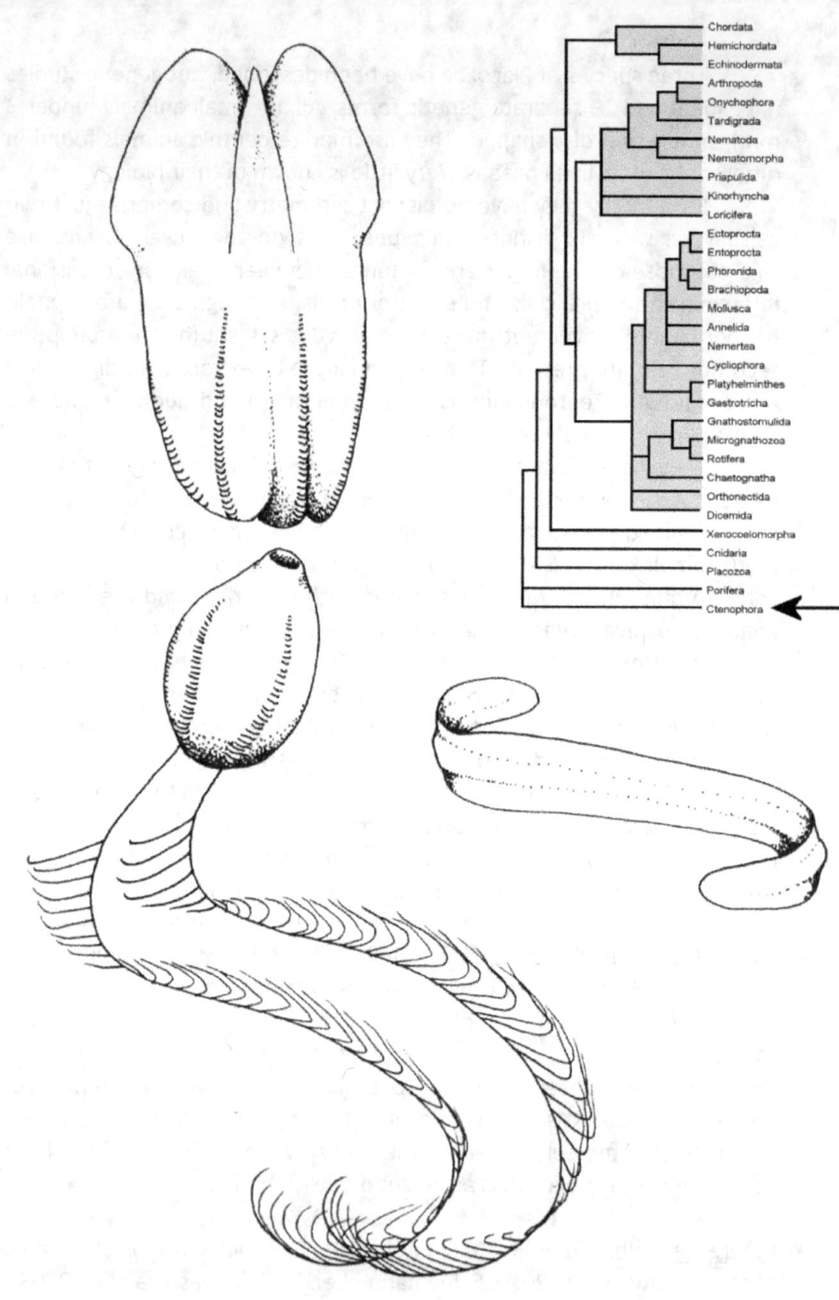

Chordata
Hemichordata
Echinodermata
Arthropoda
Onychophora
Tardigrada
Nematoda
Nematomorpha
Priapulida
Kinorhyncha
Loricifera
Ectoprocta
Entoprocta
Phoronida
Brachiopoda
Mollusca
Annelida
Nemertea
Cycliophora
Platyhelminthes
Gastrotricha
Gnathostomulida
Micrognathozoa
Rotifera
Chaetognatha
Orthonectida
Dicemida
Xenocoelomorpha
Cnidaria
Placozoa
Porifera
Ctenophora

5.   **Ctenophora** (comb-jellies)

Ctenophora are named for their characteristic ciliated 'comb-plates', the name being Greek for comb-bearing. The comb-jellies are exclusively marine pelagic predators superficially like cnidarian jellyfish (Chapter 6). They live from near the sea surface down to a depth of over 7,000 m. 250 species have been described, but the diversity and taxonomy of this group are very poorly known.

They are mostly globular, although some are ribbon-shaped. All have a unique form of rotational symmetry. Locomotion is mostly by movement of the cilia on the 'comb' plates. Some movement can also be by limited muscular contraction. The muscles are in the form of bunches of actin-rich fibres. Coordination is achieved by two nerve nets (one ectodermal and one in the mesogleal space). The nerves and muscles seem to have an independent origin from those of other animals.

All are predators and some specialise in eating urochordates or other comb-jellies. Prey are captured using adhesive and entangling colloblast cells. These can also secrete toxins. Some species also have nematocyst stinging cells; these are not produced by the comb-jelly but are obtained by consuming jellyfish tentacles. The digestive system has generally been assumed to be a bottle-gut but it has been demonstrated that one of the two small, temporary 'anal pores' functions as an anus.

Reproduction is usually sexual, although some can reproduce asexually by fragmentation. The pattern of embryonic cell division is highly unusual but is shared with Cnidaria. Development is direct; there is no larva.

Structurally they are diploblast; and their structure and general appearance led to them being classified with cnidarians in the 'Coelenterata'. Recent data indicate that they are not closely related to cnidarians, and may even be the most basal animal phylum. Their genomes are very limited in neuronal, immune or developmental genes (lacking both ParaHox and Hox). They also lack microRNAs, supporting the idea of divergence from other animals before these regulators evolved.

Fossils date from the Cambrian (around 520 million years ago).

References: Babonis et al. 2018; Jaeger et al. 2011; Lindsay & Miyake 2007; Maxwell et al. 2012; Moroz & Kohn 2016

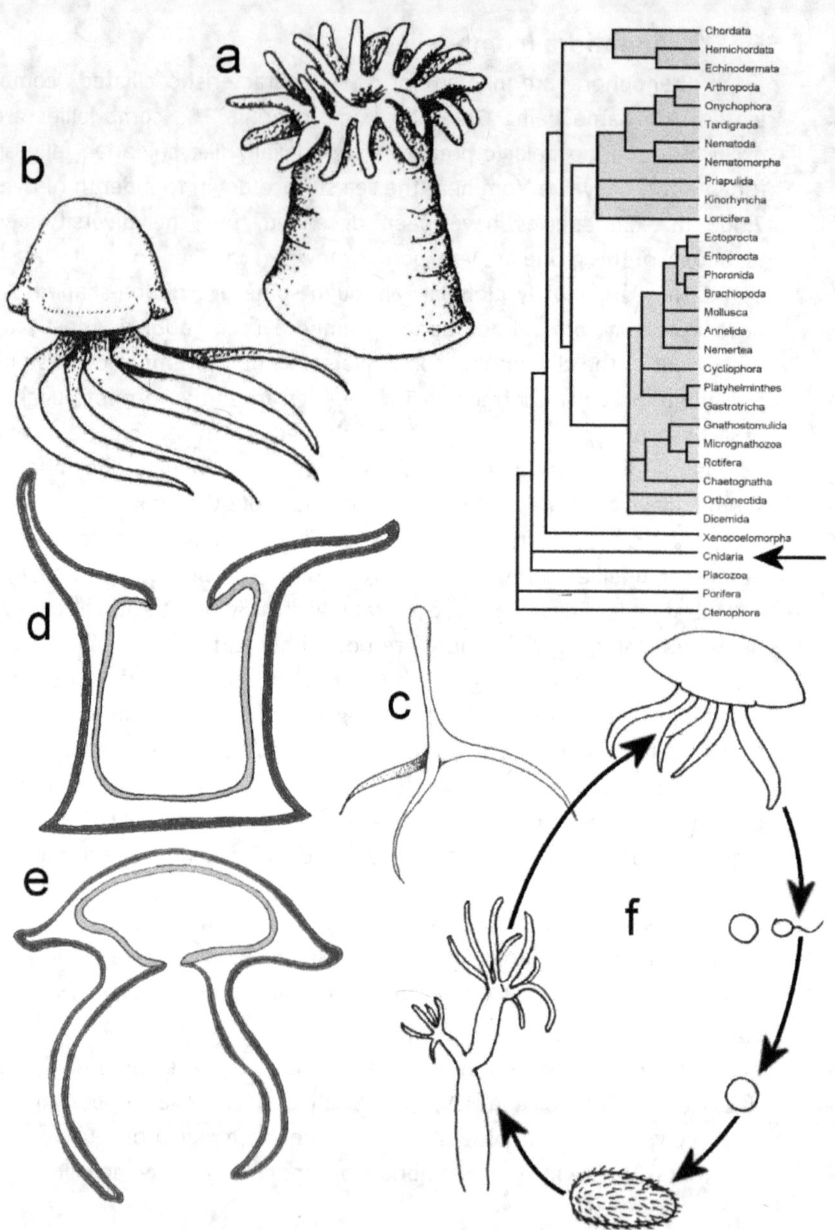

Cnidaria: a) polyp, b) medusa, c) endocnidozoan, d & e) sections showing tissue layers (see page 4), f) generalised life-cycle

6.   Cnidaria (sea anemones, jellyfish & myxozoans)

Cnidarians are named for their characteristic stinging tentacles, knide being Greek for a stinging nettle. Cnidarians are the familiar sea anemones, corals and jellyfish. All are aquatic, mostly marine predators. 12,300 species have been described in three well-supported clades: Anthozoa (sea anemones, corals), Medusozoa (jellyfish) and Endocnidozoa (*Polypodium* and Myxozoa). The latter are the 2,200 strange parasite species.

Cnidarians are diploblasts with an outer ectoderm and inner endoderm separated by a poorly defined mesoglea. They have distinct upper and lower surfaces and are radially symmetrical. The gut has only a single opening, although some deep-sea corals have modified this to a more U-shaped through-flow system.

All are predators, catching prey by means of specialised 'cnidocyte' adhesive or stinging cells. These contain a barbed projectile structure, the nematocyst, associated with a variety of toxins. Many species (corals in particular) have a symbiosis with algae (usually zooxanthellae).

A nerve net is present and muscles are mainly ectodermal. These are mostly smooth, although striated muscle is present in the swimming bell of jellyfish. This lacks the titin and troponin complexes found in the striated muscles of bilateral animals, so seems to be convergent. They have light detecting eye-spots, or a camera-eye with a lens.

Life cycles may be divided into a sessile polyp and a pelagic medusa (jellyfish). Typically the medusa is the reproductive stage, although many polyp forms are reproductive and have lost the medusa. Many jellyfish have also lost the polyp stage. Reproduction is sexual or asexual by budding, which may result in colonies. Siphonophores like the Portuguese man-o-war *Physalia physalis* are pelagic colonies of several different polyp forms. There is also often a ciliated planula larval stage.

The oddest cnidarians are the endocnidozoans, intracellular parasites of annelids, ectoprocts and vertebrates. Their intracellular nature led to them originally being thought to be protozoans. They have a two stage life-cycle and in one genus there is a free-living medusa stage.

Their well-defined but simple structure has always placed cnidarians near the base of the animal tree. They have Hox genes, but in an unusual arrangement. The fossil record extends to the Ediacaran.

References: Giribert & Edgecombe 2020; Martin 2002; Steinmetz et al. 2012

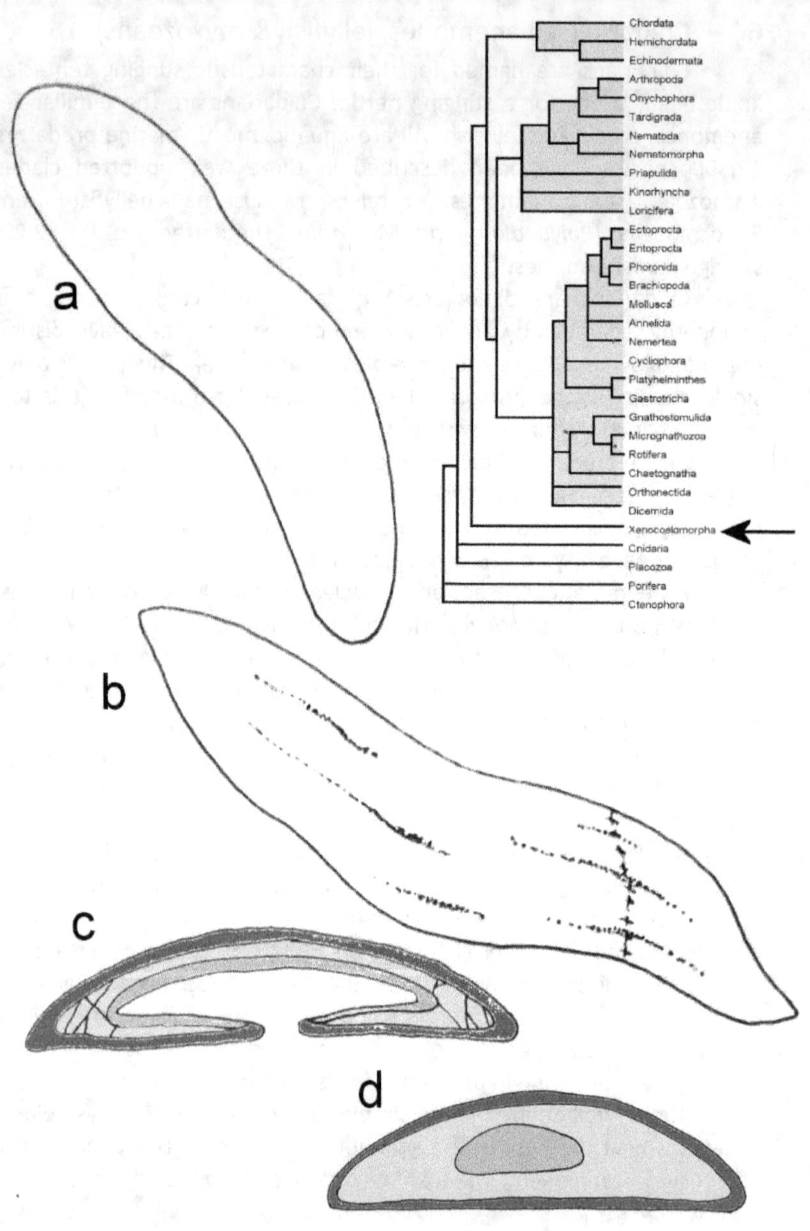

Xenacoelomorpha (sections see page 5): a & d) Acola, b & c) Xenotubellida

7. **Xenacoelomorpha** (basal flatworms: Neodermatida, Acoela, Xenoturbellida)

Xenacoelomorphs are made up of three similar largely marine flatworms, and the name is similarly a compound. Acoela and Neodermatida used to be classified as part of Platyhelminthes, the main flatworm phylum. They are genetically very different and are now placed as the basal bilaterians, along with the enigmatic Xenotubellida. Whether these are related phyla or a single phylum remains unclear.

Six Xenoturbellida species have been described from 400-1,700 m depth. Acoela comprise 400 species, all marine except for two from lakes. Some live in the sand (and are under 1 mm long), others may be active on the surface of the sea-bed, or on corals (these may reach 20 cm long). They may be predators, use symbiotic algae for nutrition or be parasitic.

All are bilaterian triploblasts, with clearly defined ectoderm, mesoderm and endoderm. They appear to be primitive in having a bottle-gut and limited brain development. Nerves may be arranged in a nerve net or concentrated into fibres. Acoels may have rhabdomeric eyes, the others are blind. Muscles are well developed, arranged into circular, longitudinal and radial fibres. The epidermis is ciliated for locomotion.

In acoels the gut is syncytial, being made up of a mass of fused cells, effectively creating a giant multinucleated cell. This fills the entire digestive system which consequently lacks a central lumen, giving rise to the name 'acoel'. In at least some species the syncytial cell takes up algae and remove the cell wall. These reduced algal cells are then passed into the rest of the body, where they form a photosynthetic body wall.

The hermaphrodite gonads are not organised into distinct tissues. Fertilisation is probably external. Embryonic development is described as unique 'duet spiral cleavage' and is direct, producing a small worm which is fully developed except that it lacks a gut.

An early molecular study placed Xenoturbellida with molluscs, but this was due to contamination with its favoured food. They were also thought to be related to Deuterostomia, but recent analyses place them with Acoela and Neodermatida.

Along with all bilaterians the Hox genes have increased to three compared to the two of cnidarians (anterior and posterior, with a central one in xenacoelomorphs). Hox genes are expressed simultaneously, rather than sequentially; this may be the primitive condition.

<u>References</u>: Bailly et al 2014 ; Giribert et al. 2016; Rouse et al 2016

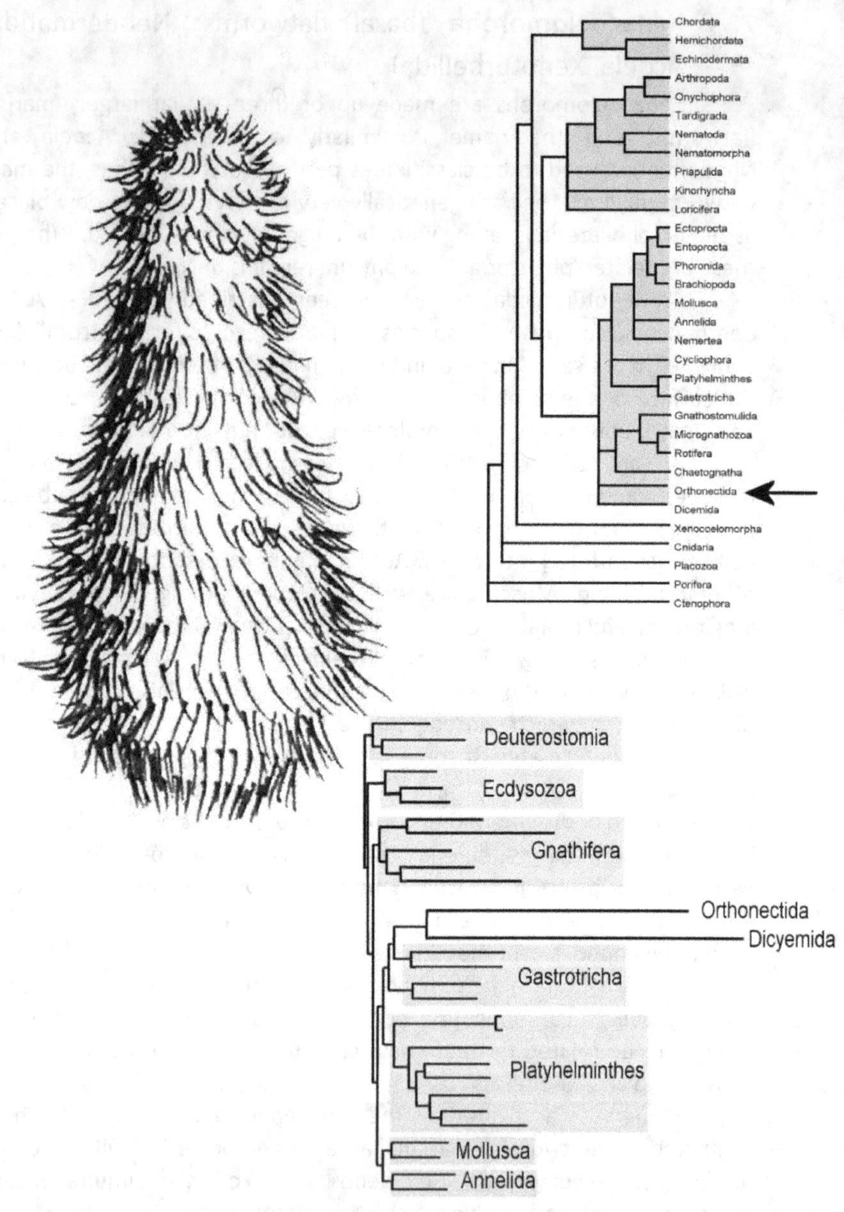

Orthonectida and phylogeny showing long-branch attraction with Dicyemida

8.   Orthonectida

Orthonectida is derived from the Greek for straight (orthos) and swimming (nektos). This appears oddly mundane, but when Giard discovered them in 1878 he was struck by their swimming "in a straight line and with the rapidity of an arrow".

Orthonectids are minute parasites of marine invertebrates. They have very simple structures but complex life-cycles. 25 species have been described, all parasitising a wide range of marine invertebrates. The free-living adult stage is short-lived and non-feeding. Mating between the 1 mm female and the microscopic dwarf male produces a larva. The larva infects a host, at which stage it loses the ciliated ectoderm and the internal cells dissociate and disperse through host cells. There they become infectious parthenogenetic 'agametes'.

Orthonectids used to be grouped with another similarly structurally simple phylum, the Dicyemida (Chapter 9), as the Mesozoa. They were thought to be intermediate between protists and animals. The relationship between 'Mesozoa' phyla is not clear due to their simplified morphology and to long-branch attraction where highly divergent molecular data sets group together. Early molecular studies supported their basal or even pre-animal position but more recent studies have placed them in the Protostomia (specifically in the Spiralia).

Protostomes are defined by an embryonic development that supposedly differs from that of deuterostomes (Chapter 31). The main embryological differences are the pattern of cleavage (early cell division), mouth and anus origin, mesoderm and coelom formation. Protostome cleavage is generally spiral whereas that of deuterostomes is radial. Whilst this holds for deuterostomes it is not an accurate division for protostomes and the fully spiral pattern is only associated with one of the subdivisions of the protostomes, the Spiralia. In most protostomes, the original infolding of the embryo, the blastopore, tends to form the mouth, but this is not consistent and in some it forms the anus in the deuterostome manner. Mesoderm forms from the edge of the blastopore and the coelom as a split in this mesoderm in protostomes, or from the developing gut with a pocket from the gut forming the coelom in all deuterostomes. Again, there are exceptions. Despite this unreliable developmental division, protostomes and deuterostomes are supported as monophyletic groups by molecular phylogenies.

<u>References</u>: Lu et al. 2017; Schiffer et al. 2018

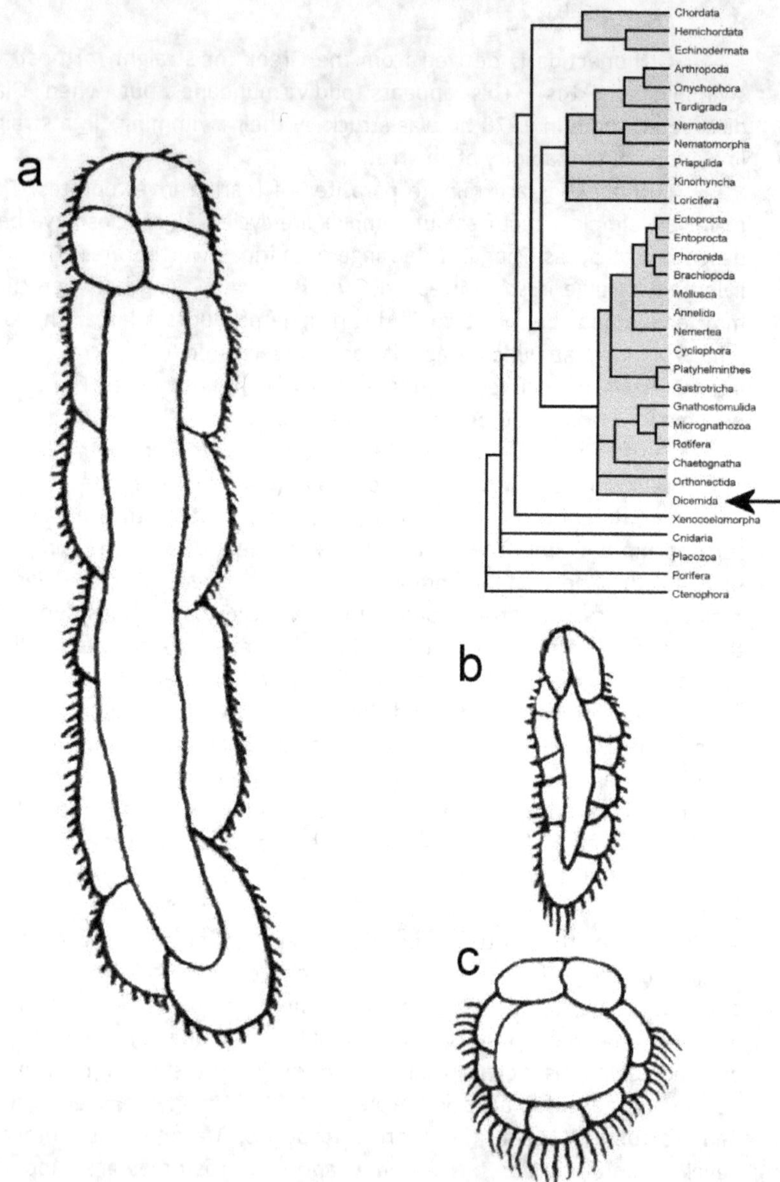

Dicyemida: a) adult, b) vermiform embryo, c) infusoriform embryo

9.   Dicyemida

Dicyemida (or Rhombozoa) are a group of parasites in the kidneys of octopuses and cuttlefish. They are mostly microscopic, and even those that reach 7 mm in length consist of a maximum of just 41 cells. They used to be grouped with another similarly structurally simple phylum, the Orthonectida (Chapter 8), as the Mesozoa.

124 species have been described, but the number of potential host species suggest that there may be closer to 500 species in total. As they are host specific their distribution is directly linked to their hosts, meaning that they are an entirely marine group found from the surface to 4,700 m depth.

These are extremely simplified animals, lacking tissues and any form of symmetry. Their cells lack a basement membrane but they do produce its components, including fibronectin, laminin and type IV collagen. The presence of these components indicates secondary loss of the basement membrane. Their secondary simplification is so extreme that they converge on protists in having tubular (rather than lamellar) mitochondrial cristae, endocytosis (taking material into a cell by invagination of the membrane to form a vacuole) and an absence of collagen. The mitochondrial cristae form by invaginations of the inner membrane, but the significance of tubular versus lamellar ones is obscure. The endocytosis is presumably the result of the loss of tissue complexity resulting in a protist-like digestion.

The worm-shaped adult has a large axial cell surrounded by 14-40 ciliated cells. Within the central polyploid cell of the adult an asexual reproductive cell, the 'agamete', develops. This may give rise to two different organisms: the vermiform embryo or the 'infusorigen' stage. The vermiform embryo develops directly into the adult, whilst the 'infusorigen' reproduces asexually to produce the gametes. These fuse to give the 'infusoriform' embryo. These infect the next host, although how they do this is not known. This production of two different sorts of embryo is the trait that gives the phylum its name (di-, plus Greek from embryo – kyema).

As with the Orthonectida, their phylogenetic position has been uncertain. Now they are regarded as close to the Platyhelminehtes (Chapter 15) and Gastrotricha (Chapter 14) spiralians.

References: Furuya & Tsuneki 2007; Hochbert & Hofrichter 2005

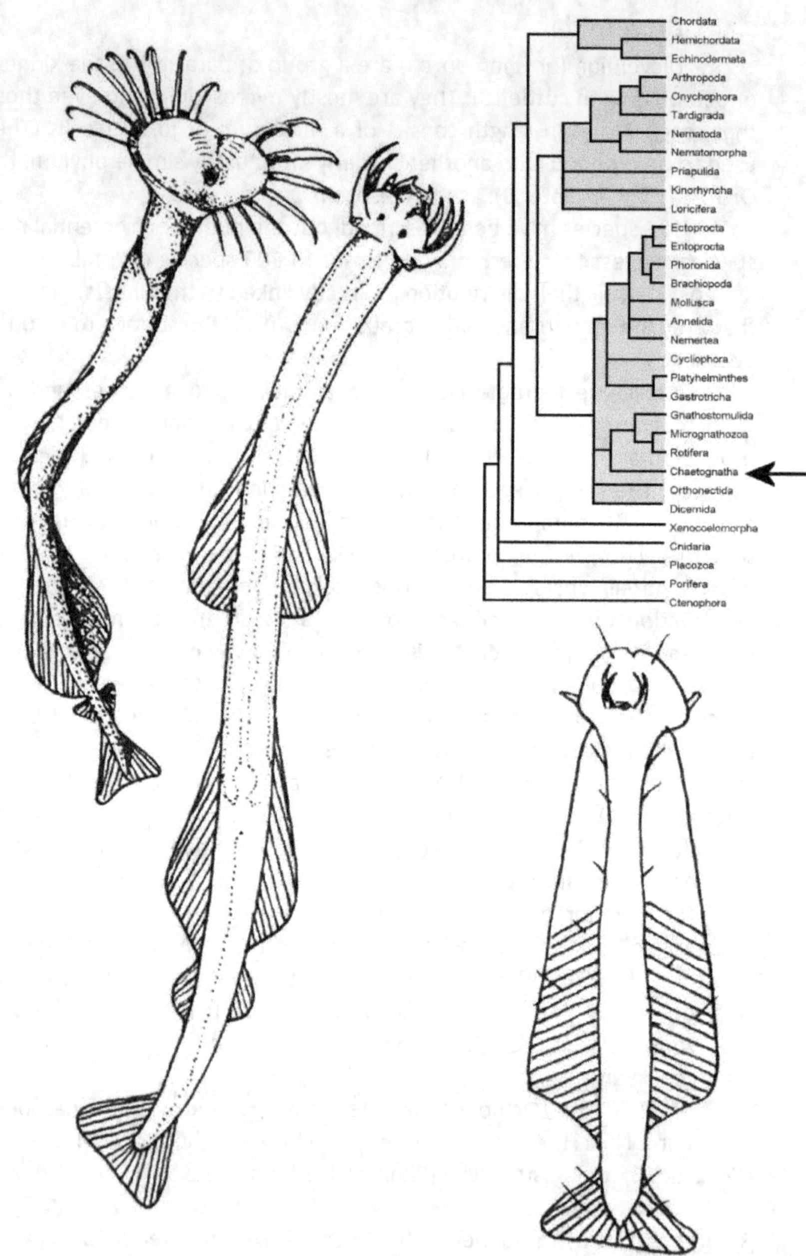

Chordata
Hemichordata
Echinodermata
Arthropoda
Onychophora
Tardigrada
Nematoda
Nematomorpha
Priapulida
Kinorhyncha
Loricifera
Ectoprocta
Entoprocta
Phoronida
Brachiopoda
Mollusca
Annelida
Nemertea
Cycliophora
Platyhelminthes
Gastrotricha
Gnathostomulida
Micrognathozoa
Rotifera
Chaetognatha
Orthonectida
Dicemida
Xenocoelomorpha
Cnidaria
Placozoa
Porifera
Ctenophora

10.   Chaetognatha (arrow worms)

Chaetognaths are the most important phylum that most people have never heard of. They are planktonic predators, making up 10% of the plankton. Hence they are the most abundant predators on earth and play a major role in marine environments and thus in global ecosystems.

Most of the 190 species are 1-2 mm long but a few species reach 12 cm. They are found from the surface to a depth of 2,500 m. All have a similar form: elongate worms with a head, trunk and post-anal tail divided by septa. They have distinctive lateral and caudal fins supported by rays. The fins are used in gliding, after which they use their longitudinal muscles (there are no circular muscles) for active swimming upwards.

They prey on copepod crustaceans using the complex jaws. They have a set of grasping spines and teeth (which give the phylum its scientific name – spine + jaw, Greek khaite + gnathos), covered by a retractable hood which acts as a net. Prey are trapped in the net and subdued with tetrodotoxin sequestered from the alga *Vibrio alginolyticus*. Prey are detected with sensory setae and some vision produced by compound pigment-cup ocelli. The brain is in the form of dorsal and ventral ganglia. Gas exchange, excretion and circulation rely on diffusion.

Arrow worms are hermaphrodites and have simultaneous fertilisation, each placing a spermatophore on the neck of their mate, this releases sperm which swim along the body to the female pores at the base of the tail. The eggs are planktonic or attached to algae. Development is direct; no larval stage is needed in planktonic animals.

Arrow worms have always been a phylogenetic puzzle. They have been suggested to be related to many different groups, including the deuterostomes. However more comprehensive studies suggest a relationship to gnathiferans (sharing grasping spines). The deuterostome similarity is merely superficial; the blastopore appears to form the anus but in fact closes early with a new anus forming later and the cells around it do not express the brachyury gene. This T-box transcription factor is most notably expressed in mesoderm of chordates (Chapter 34). In other phyla it is expressed around the blastopore, that it is not expressed in this region in arrow worms shows that this is not genuine deuterostomy.

Fossils date from 520 million years ago, with slightly earlier microfossils spines 'protoconodonts' (535 million years ago).

References: Ball & Miller 2006; Laumer et al. 2019; Szaniawski 2002; Takada et al. 2002

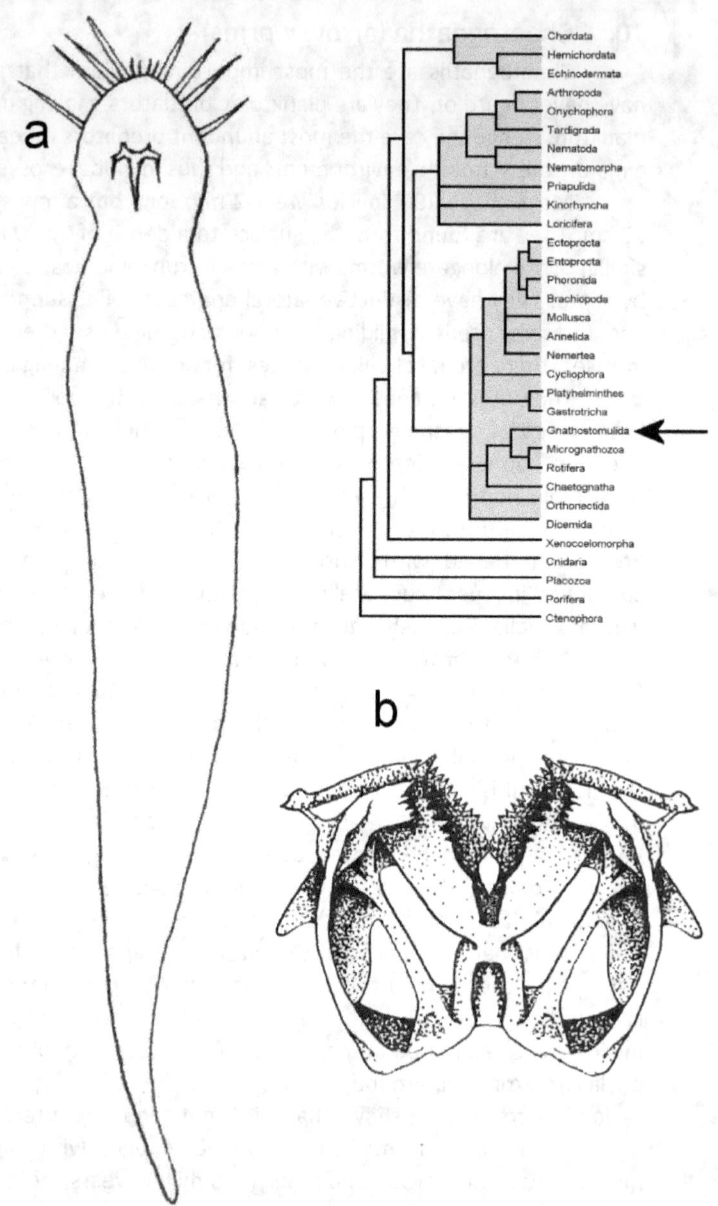

Gnathostomulida: a) morphology, b) gnathiferan jaws

11.   **Gnathosumulida** (spiny jaws)

The Gnathostomulida are named for their only conspicuous feature: the spiny jaws, being derived from the Greek gnathos, jaws, and stoma, mouth.

Gnathostomulids are 100 species of minute marine animals living in the meiofauna, with the exception of one brackish water species. They feed on bacteria in detritus rich, high sulfide, low oxygen sediments at 30-400 m below the surface. They range in size from 0.3-3.6 mm.

They were first discovered in 1928 but not described until 1956, originally as a form of flatworm. They were recognised as a phylum in 1969. They have been considered to be related to gastrotrichs, ecdysozoans, chaetognaths and more recently to rotifers. Gnathostomulids are grouped along with rotifers and micrognathozoans in the Gnathifera. They all share a similar jaw structure of a series of rods and articulating pincers. Hox genes suggest that chaetognaths might be part of the gnathiferans, which also have high levels of chitin in the teeth. There is some suggestion that cycliophorans might group here as well.

Gnathiferans have a body with two regions: a head and trunk. They are structurally simple, being unsegmented and acoelomate. They are ciliated (along with free-living platyhelminthes, nemerteans, gastrotrichs and many larvae). The brain is represented only by a buccal ganglion. There is no circulatory system or gas exchange structures. The gut is fully developed but is a bottle-gut. One genus does have a temporary opening on the dorsal side, which might function as an anus.

Reproduction is sexual and all known species are hermaphrodites, presumably with internal fertilisation but this has not been observed. Development is direct.

There are no unambiguous gnathiferan fossils due to their small size but there have been some finds of what look like gnathiferan jaws. These jaws led to the identification of *Amiskwia* of the Burgess Shale (505 million years ago) as a possible gnathostomulid.

References: Caron & Cheung 2019 ; Laumer et al. 2019; Sorensen et al. 2000

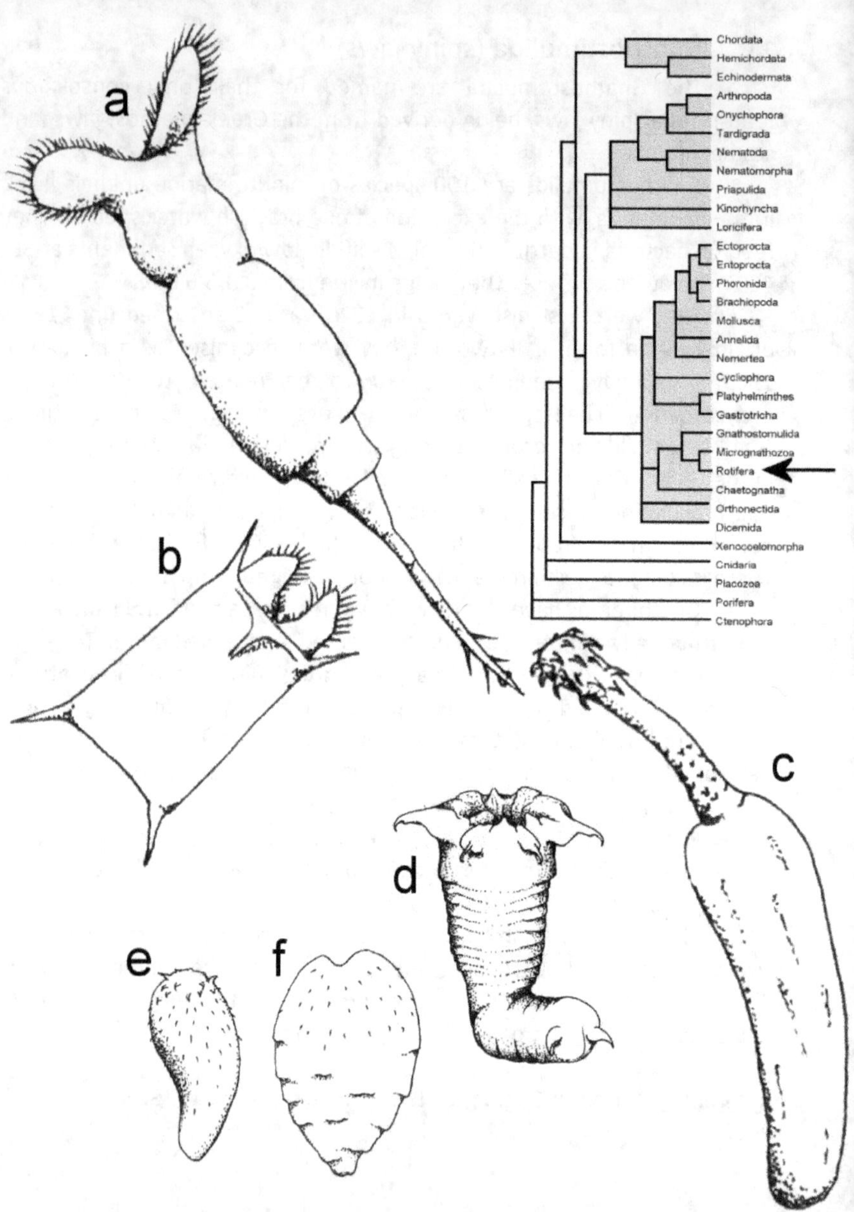

Rotifera: a) Bdelloidea, b) Monogononta, c) Acanthocephala, d) acanthor larva, e) acanthella larva, f) cystacanth

12. **Rotifera** (wheel animals and spiny-headed worms)

Rotifers comprise two distinct forms: free-living wheel animals (the original Rotifera group) and the parasitic spiny-headed acanthocephalans which were once thought to be a distinct phylum. Over 2,000 species of free-living rotifers have been described and 1,200 acanthocephalans.

Wheel animals are minute (to 1 mm) but acanthocephalans range from 1 to 60 cm in length. They are eutelic, meaning that no cell division occurs in adults. This oddity is found in other small organisms: many nematodes, tardigrades and dicyemidans. Wheel animals are aquatic predators, using a ciliated crown (the wheel organ that gives the group its common and its Latin name) for prey capture and locomotion. The jaws are in the form of a chewing pharynx ('mastax') armed with calcified jaws ('trophi'). They are found in freshwater, damp soil and subterranean water to 1.4 km underground. All are solitary but in 25 species aggregations of up to 3,500 individuals can occur, resembling colonies.

The 460 species of bdelloid wheel animals are notable for having been exclusively asexual for the past 25 million years. Other groups are fully sexual or are cyclic parthenogens. Wheel animals are able to survive complete dehydration. This anhydrobiosis does not involve accumulation of trehalose sugars as in other animals able to do this (e.g. tardigrades: Chapter 29); they seem to use hydrophilic proteins instead.

Spiny-headed worms have a proboscis armed with hooks. They lack a gut, using the hooked proboscis to lodge in the host gut. They may be described as pseudosegmented but the 'segments' are just ectodermal folds. They have separate sexes. Eggs are taken up by an intermediate arthropod host, inside which they hatch into spindle-shaped acanthor larvae. These move into the body cavity, developing into acanthella larvae then cystacanths (adult worms). Consumption of the intermediate host leads to infection of the definitive vertebrate host.

The spiny-headed worms are related to the four sesionoid wheel animals which are ectoparasites on the crustacean *Nebalia*. Wheel animals have been suggested to be paedomorphic, effectively being trochophore larvae (Chapter 16) that have become sexually mature without metamorphosis into the 'adult' form. However, similarities between rotifers and trochophores are only superficial.

Fossil rotifers are known from the Eocene (40 million years ago).

References: Borgonie et al. 2015; Frobius & Funch 2017; Ricci 2016; Tunnacliffe et al. 2005

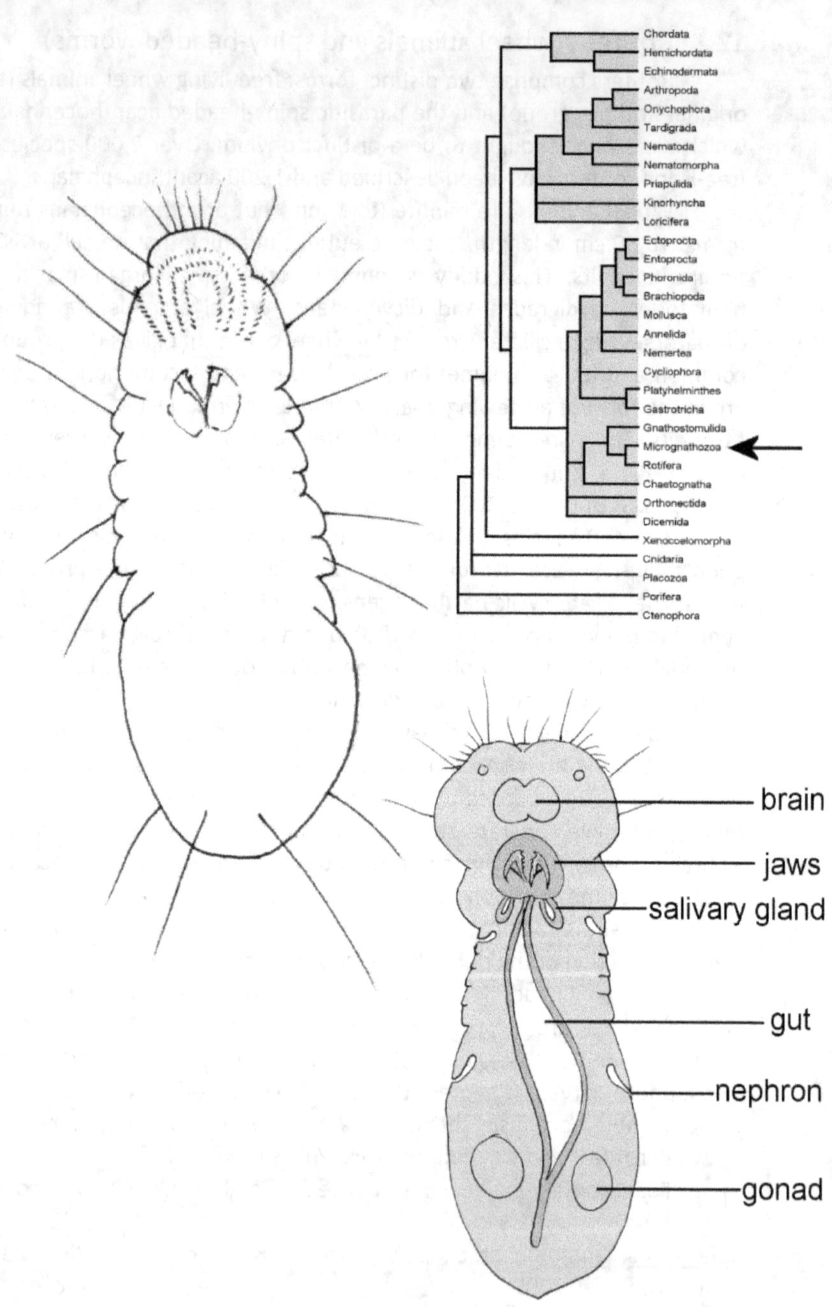

Chordata
Hemichordata
Echinodermata
Arthropoda
Onychophora
Tardigrada
Nematoda
Nematomorpha
Priapulida
Kinorhyncha
Loricifera
Ectoprocta
Entoprocta
Phoronida
Brachiopoda
Mollusca
Annelida
Nemertea
Cycliophora
Platyhelminthes
Gastrotricha
Gnathostomulida
Micrognathozoa
Rotifera
Chaetognatha
Orthonectida
Dicemida
Xenocoelomorpha
Cnidaria
Placozoa
Porifera
Ctenophora
brain
jaws
salivary gland
gut
nephron
gonad

13. Micrognathozoa

Micrognathozoans are exceptionally obscure and are certainly the least well known of all animal phyla. The first one was found in a spring on Greenland in 1995. It was described in 2000 and recognised as a distinct phylum in 2004. *Limnognathia maerski* remains the only species to have been described but what is probably a second species has been collected from Crozet islands near Antarctica. It is uncertain whether specimens from Wales and England belong to the named species.

These minute animals (150 microns) have little in the way of structure and their names refer to the only conspicuous feature: the jaws. That is in the form of Micrognathozoa – the tiny (micro) jaws (gnathos) animal (zoa), and *Limnognathia* – the freshwater jaw. They are unsegmented, acoelomate, lacking circulation or gas exchange structures. The gut is not fully developed, with only a temporary anus. The body can be divided into a head, a compressible thorax and an abdomen. Within the head there is a complex jaw apparatus with four sets of elements which are used in predation on bacteria. The brain is a slightly bilobed dorsal structure with paired ganglia below the oesophagus.

They have a distinctive locomotion when moving in water, with a slow, spiral motion generated by cilia. They can also glide or crawl over a surface.

Only females have been found, so the phylum is probably parthenogenetic. Development is direct, with no larval stage.

Due to their microscopic size there is no fossil record.

References: Giribert et al. 2004; Kristensen & Funch 2000; Worsaae & Kristensen 2016

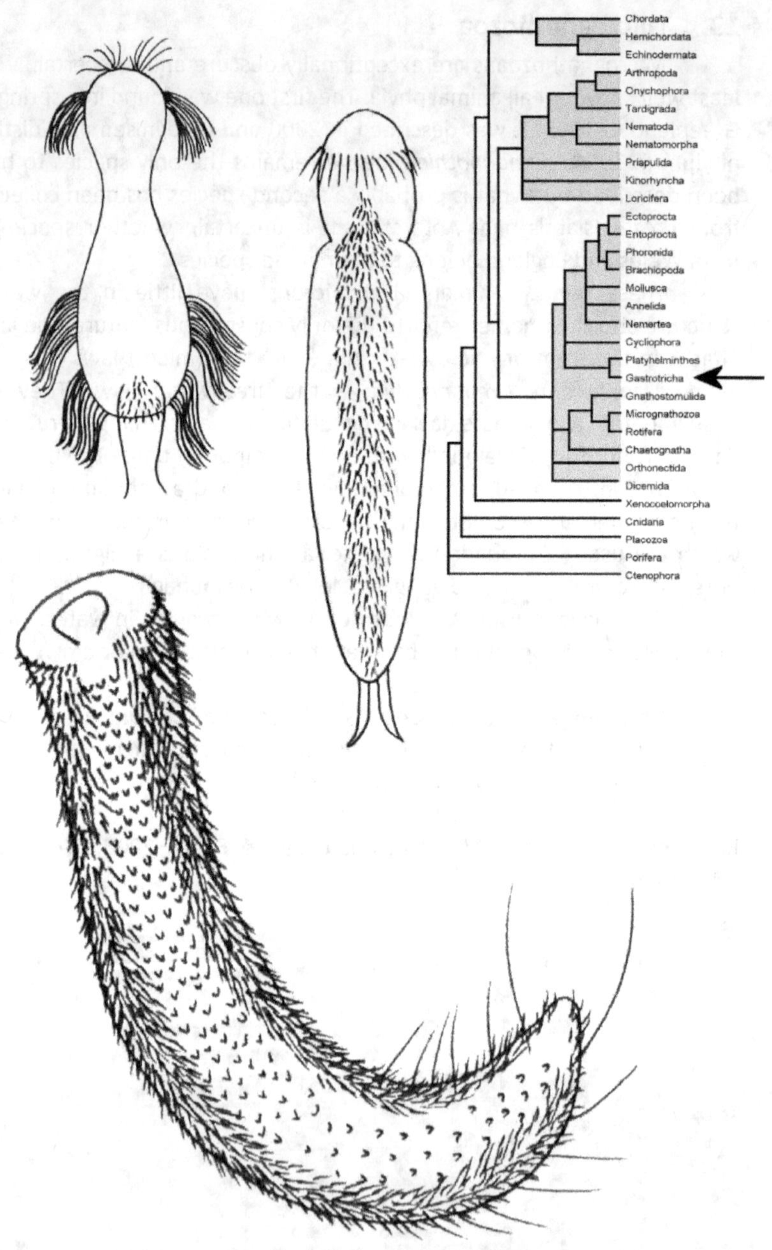

Chordata
Hemichordata
Echinodermata
Arthropoda
Onychophora
Tardigrada
Nematoda
Nematomorpha
Priapulida
Kinorhyncha
Loricifera
Ectoprocta
Entoprocta
Phoronida
Brachiopoda
Mollusca
Annelida
Nemertea
Cycliophora
Platyhelminthes
Gastrotricha
Gnathostomulida
Micrognathozoa
Rotifera
Chaetognatha
Orthonectida
Dicemida
Xenocoelomorpha
Cnidaria
Placozoa
Porifera
Ctenophora

14.   Gastrotricha (hairybellies)

Gastrotricha have a characteristic appearance, being covered in hairs; hence the scientific and common names.

Gastrotrichs are microscopic predators, mostly living between sand grains in wet environments (marine, freshwater and terrestrial) although some are planktonic. Interstitial species living in sand may be locally abundant and are found from the subtidal zone to a depth of 5,700 m. They feed on detritus, bacteria and algae.

The largest of the 840 described species reach 3.5 mm, but most are under 1 mm long. Superficially they look like flatworms, being unsegmented acoelomates with cilia on the ventral side and lacking gas exchange and circulatory systems. Hairybellies have been speculated to be related to rotifers, kinorhynchs, nematodes and gnathostomulids. They are now thought to be with platyhelminths in the Rouphozoa, but this remains uncertain. The most obvious external difference between gastrotrichs and platyhelminths is the addition of adhesive tubes at the end of the body and a thickened cuticle covered in plates and spines. In addition to conventional longitudinal and circular muscle they have a unique helicoid muscle system around the gut.

The brain is in the form of bilaterally symmetrical ganglia on either side of the head, bridged by dorsal and ventral commissures. There is very limited vision using pigmented eyespots.

They are all sexual hermaphrodites although some may have a parthenogenetic stage. They have simultaneous reciprocal internal fertilisation where the sperm is transferred from the male pore to the caudal organ which functions as a penis. Development is direct.

Due to their microscopic size there is no fossil record.

<u>References</u>: Hochberg & Litvaitis 2001; Marletaz et al. 2019; Schmidt & Martinez Arbizu 2015

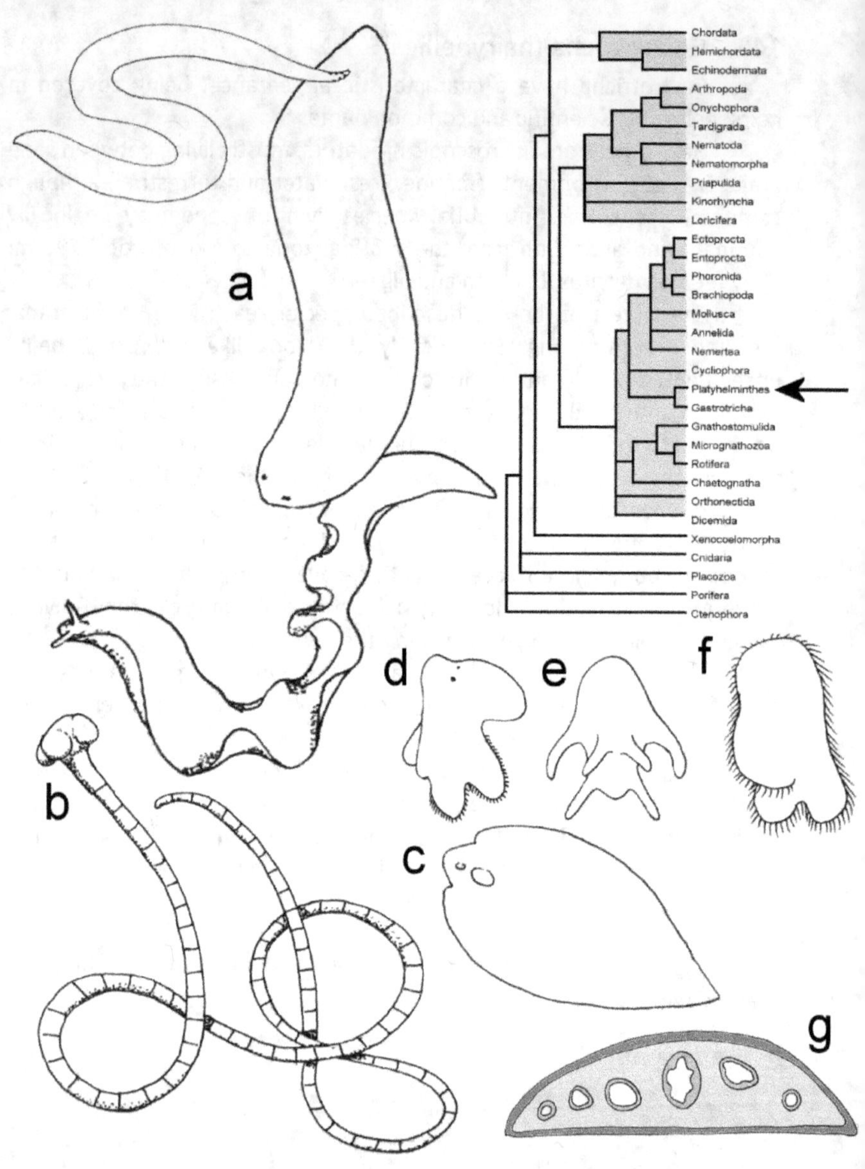

Platyhelminthes: a) free-living, b) tapeworm, c) fluke, d) Müller's larva,
e) Goette's larva, f) Kato larva, g) section (see page 4)

15. **Platyhelminthes** (flatworms)

Platyhelminths are the true flatworms, until recently thought to include the Acoela and Neodermatida, which are now regarded as basal bilaterians (Chapter 7). There are 30,000 marine, freshwater, terrestrial and parasitic flatworms. 75% of described species are parasites (11,000 fluke species, 6,000 tapeworms and around 5,000 less specialised forms), some species of major medical and agricultural significance. Non-parasitic forms are almost all predators, with a few scavengers and very few herbivores. They are found in wet environments down to a depth of 9 km in the sea and 1.4 km in ground water. Most are very small, less than a millimetre, free-living species may reach 1 m, parasites to 8.8 m.

Flatworms have exceptional regeneration capabilities and are well known for their ability to regrow missing body parts. Differentiated cells have no mitosis but throughout their lives the flatworms retain stem cells that migrate throughout the body and repair damage. This also results in the regeneration, which some use deliberately as asexual reproduction.

Locomotion is by a mixture of ciliary and muscular action. They are acoelomate triploblasts, with an ectoderm, a basic mesoderm and an endoderm. The gut is a blind-ended bottle-gut, although in one genus an anal pore is present. In some parasites the gut may be absent entirely. Where present, the gut branches throughout the body, ensuring that diffusion can operate between the digestive system and all cells. Most processes rely on diffusion, as they lack circulation and specialised gas exchange. Excretion uses flame cells (also seen in nemerteans: Chapter 19). The brain is an anterior cerebral ganglion. They have cup-shaped ocelli of rhabdomeric photoreceptors and pigment cells.

Predators use toxins to subdue their prey, including tetrodotoxin. Some can sequester nematocysts from their prey. Defensively they can produce acid secretions and some have calcareous spicules.

They are usually internally fertilising hermaphrodites. Aquatic forms may produce a Goette's, Müller's or Kato's larva of increasing complexity. Parasitic forms often have very complex lifestyles.

The fossil record is unclear; eggs or hooks of parasites may be present in Devonian fish. A definite flatworm is present in Eocene amber.

References: Giribert & Edgecombe 2020; Stokes et al. 2014

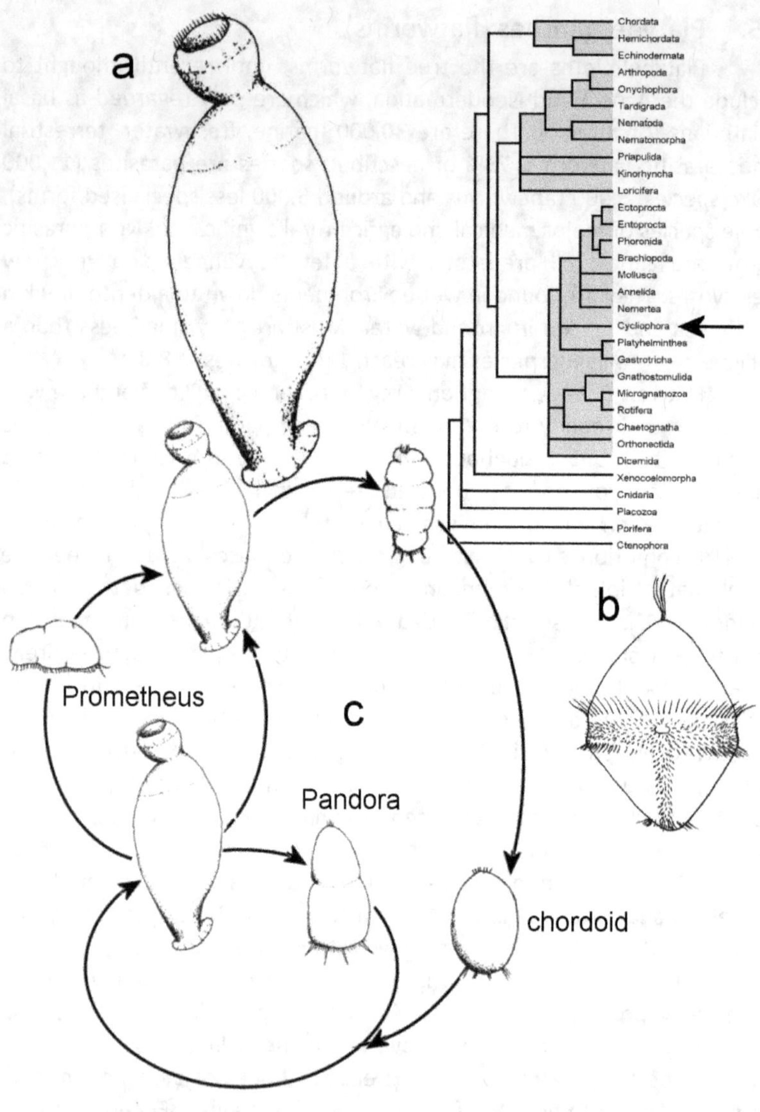

Cycliophora: a) morphology, b) life-cycle c) lophotrochozoan trochophore

16. Cycliophora

Cycliophora have the narrowest niche of any phylum found to date. They were first described in 1995 from specimens on the mouthparts of lobsters. Four species have been found on three different lobster species. The phylum is named as the 'circle bearers' (Greek kuklios + pherein) in reference to the apical tentacular feeding structure.

They have little in common with any animal, and the complex life-cycle was originally thought to suggest a relationship to Ectoprocta (Chapter 20). Molecular analyses have been more useful, suggesting a link to rotifers or ectoprocts. This seems to be the most stable solution, all grouping at the base of the Lophotrochozoa. The Lophotrochozoa were proposed in 1995 to unite the lophophorate phyla (Brachiopoda, Ectoprocta and Phoronida), with the molluscs and annelids. It was later expanded considerably and is now largely synonymous with Spiralia (excluding Gnathifera and Rouphozoa). As the name implies, its members either have a lophophore feeding apparatus or some form of 'trochophore' larva. The trochophore has a preoral band of locomotory cilia and postoral and circumoral feeding bands.

Cycliophorans are exceptionally small, under 500 microns. Being so small diffusion is sufficient for most transport needs and accordingly they are unsegmented acoelomates, lacking circulation or gas exchange structures. The body has a ciliated buccal funnel, an oval trunk and a posterior adhesive disc. This sticks the animal onto the setae of the mouthparts of lobsters where food particles are drawn into the U-shaped gut through the ciliary action on the funnel.

The life-cycle has alternating sexual and asexual stages with unique details. Several individuals develop within the asexual stage and escape either as an asexual 'Pandora larva' or a sexual individual. The Pandora larva settles on the same host and develops into a new asexual individual. In the sexual version, the escaping larva is a 'Prometheus larva'. When this settles on a feeding cycliophoran it develops males within it. These will mate with any females on the same host. The mated female then moves onto the same lobster and encysts around a fertilised egg. Finally a chordoid larva hatches from the cyst and migrates to a new host. There it restarts the asexual phase. The fully developed male is the most reduced of all animals having only 50 cells and measuring just 40 microns.

References: Funch & Kristensen 1995; Funch & Neves 2018; Marletaz et al. 2019; Passamaneck & Halanych 2006

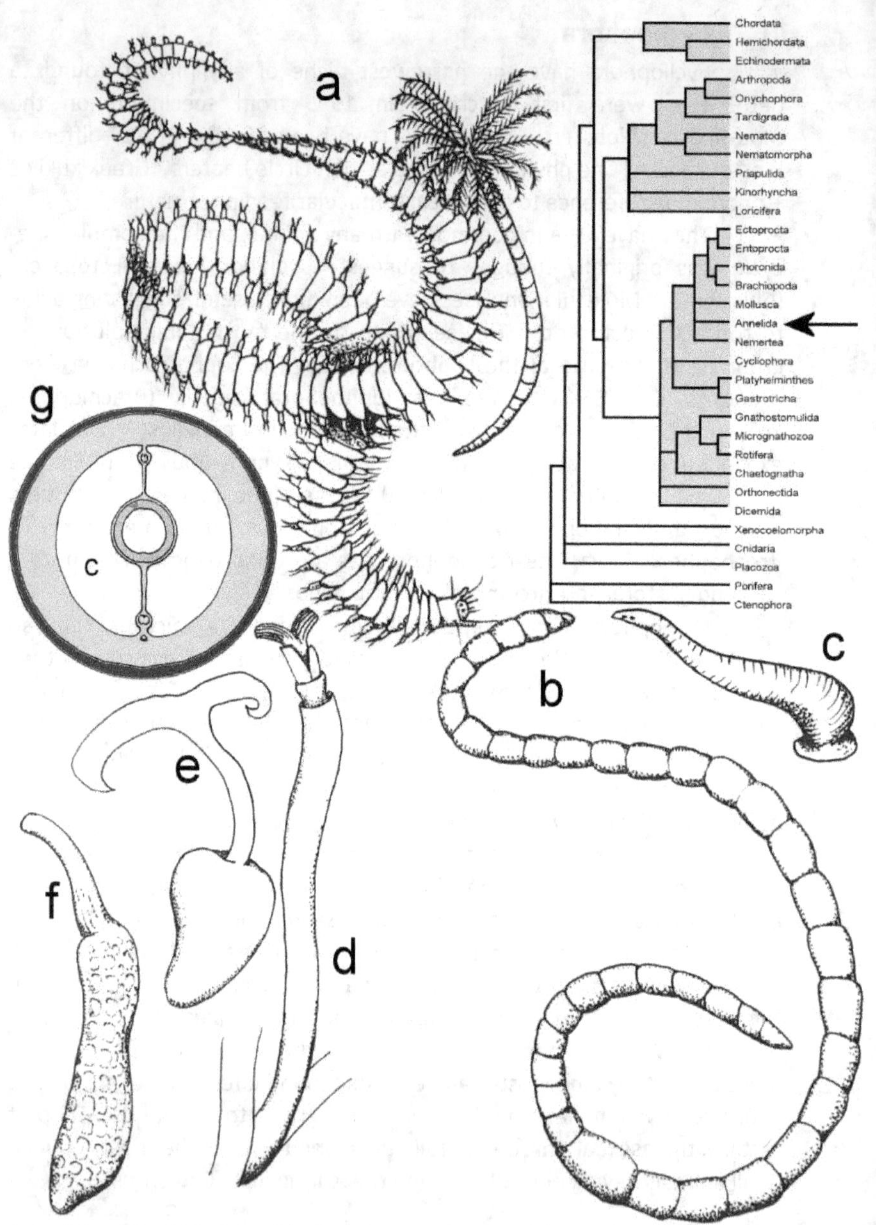

Annelida: a) polychaetes, b) oligochaete, c) leech, d) beard-worm,
e) spoon-worm, f) peanut-worm, g) section (see page 4)

17.   Annelida (segmented worms)

Annelids are the segmented worms (the name being modern Latin for ringed): the familiar terrestrial earthworms, the marine ragworms, lugworms and fan-worms, and the leeches. There are 23,000 species in marine, freshwater and terrestrial habitats. Many are very small, around 2 mm, but some can be very large; terrestrial giant earthworms may reach 2.9 m and the marine bobbit worm 6 m.

The body can be divided into a head, segmented body and terminal pygidium. The 'head' comprises the first segment (prostomium) and the peristomium (containing the mouth), and may carry feeding appendages. The segments bear chaetae in almost all groups, and many marine species have parapodia used in swimming or in pumping water.

With a few exceptions they are coelomate. Within the coelom fluid is moved by cilia or muscular contraction. There is also a closed circulatory system of blood vessels and basic 'hearts'. The respiratory pigments are varied, usually including haemoglobin and sometimes haemerythrins. The gut is lost in some extremophiles, which rely on bacterial symbiosis for nutrition instead (e.g. pogonophorans). Osmoregulatory systems are sophisticated, with well-developed nephridia. The nervous system is well organised with a brain connecting to paired ganglionated ventral nerve cords. There is limited vision with pigment cup ocelli.

They are hermaphrodite or with separate sexes. Development is direct or, in most marine species, with a trochophore larva.

Several groups were formerly thought to be separate phyla but are now placed in the Annelida: Sipuncula (peanut worms), Pogonophora (beard worms), Vestimentifera (tube worms), Echiura (spoon worms), Myzostomida, and possibly Orthonectida (Chapter 8). These are identified as annelids primarily on the basis of molecular phylogenies, but there are some morphological or developmental similarities, e.g. echiurans have a trace of segmentation in the nervous system. These old phyla are mostly very specialised: pogonophorans and vestimentiferans as the hydrothermal vent tube worms, echiurans as detritus feeders, and myzostomids as parasites of echinoderms and sea anemones.

Fossils are generally ambiguous, but reasonably convincing fossils date from the early Cambrian (around 530 million years ago).

References: Giribert & Edgecombe 2020; Rouse 2016

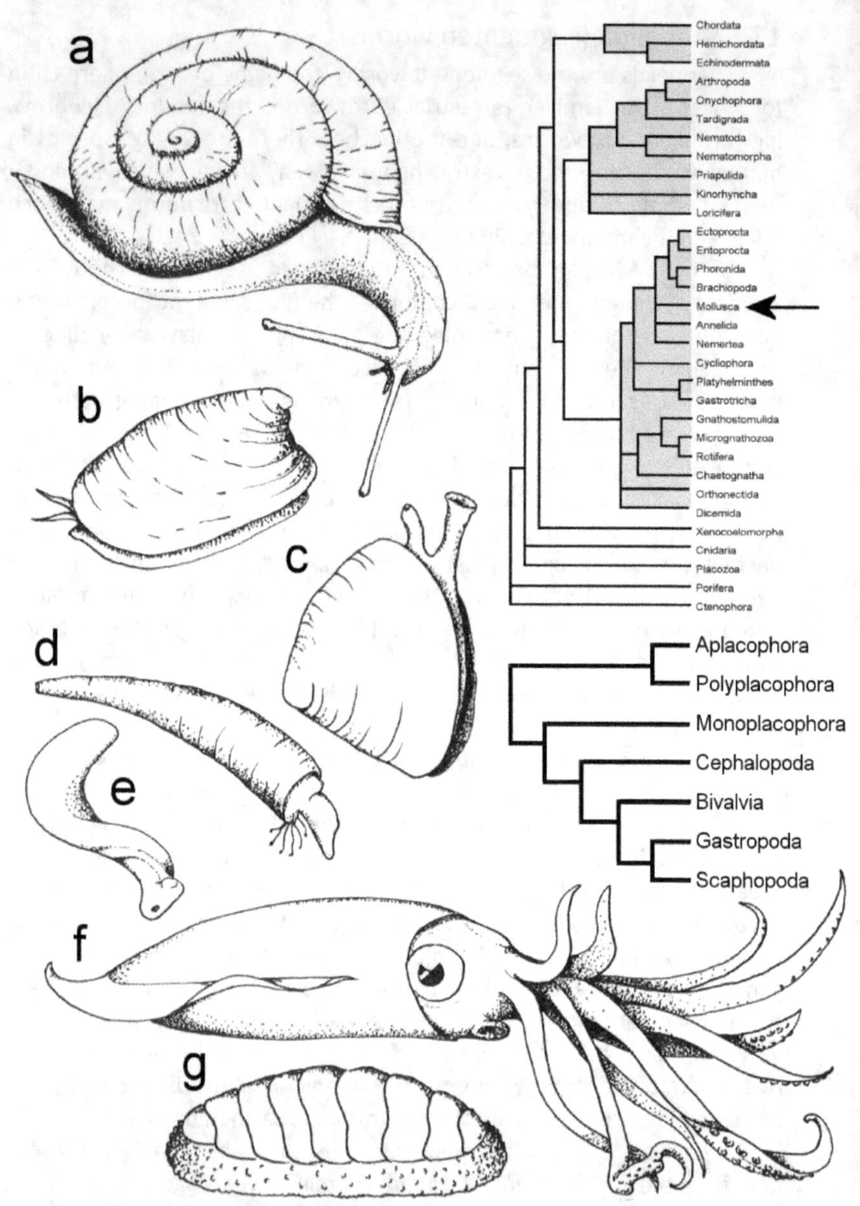

Mollusca: a) Gastropoda, b) Monoplacophora, c) Bivalvia, d) Scaphopoda, e) Aplacophora, f) Cephalopoda, g) Polyplacophora

18.   Mollusca

Linnaeus originally used Mollusca to mean a soft-bodied invertebrate (from modern Latin mollis – soft). Since then molluscs have been restricted to the familiar snails and slugs (Gastropoda), clams (Bivalvia), squid, octopus, cuttlefish and nautilus (Cephalopoda), and the more obscure tusk-shells (Scaphopoda), chitons (Polyplacophora), worm snails (Aplacophoroa) and Monoplacophora. The latter is the most obscure of all, having been known only from early Palaeozoic fossils until living animals were discovered in 1952 at 3,590 m depth.

They are found in all environments and are one of the most diverse phyla. Around 46,000 marine species have been described, but there are probably 100,000 more. 35,000 terrestrial and 1,500 freshwater species are known. They range from 0.8 mm snails to giant squid over 12 m long.

Molluscs are defined by the possession of a creeping muscular foot, a calcium shell and a feeding apparatus of an odontophore cartilage underlying a radula ribbon with microscopic teeth. Many characters have been lost in some groups. The epidermal mantle secrets calcium carbonate spicules, plates or shells, and acts as gills or a lung. This respiratory surface is supplied with haemocyanin by an open circulation and a multi-chambered heart. The gut is well developed, as befits a largely herbivorous phylum. The phylum also includes detritivores, filter-feeders and predators catching everything from nematodes to whales. The predatory cephalopods have the largest and most complex brains outside of vertebrates, and excellent vision with a vertebrate-like eye.

In addition to the mineralised shell, molluscs may have secondary hardening of the radular teeth: magnetite and other iron oxides in chitons and goethite in limpets. This enables them to graze encrusting algae off rocks without excessive tooth wear. The scaly-foot gastropod *Chrysomallon* of the hydrothermal vents has a layer of iron or zinc sulfide on the shell and a foot armoured with metal-mineralised sclerites.

Fertilisation is usually internal. Marine gastropods and all bivalves have larvae, initially a form of trochophore then a characteristic 'veliger'. The most striking aspect of development is the torsion seen in gastropods where the embryo twists around to produce an asymmetrically coiled adult. Hox genes have lost collinearity in gastropods and cephalopods.

Mollusc shell fossils date back to 540 million years ago.

References: Chen et al. 2015; Lemche 1957; Bouchet et al. 2016; Nishiguchi & Mapes 2008

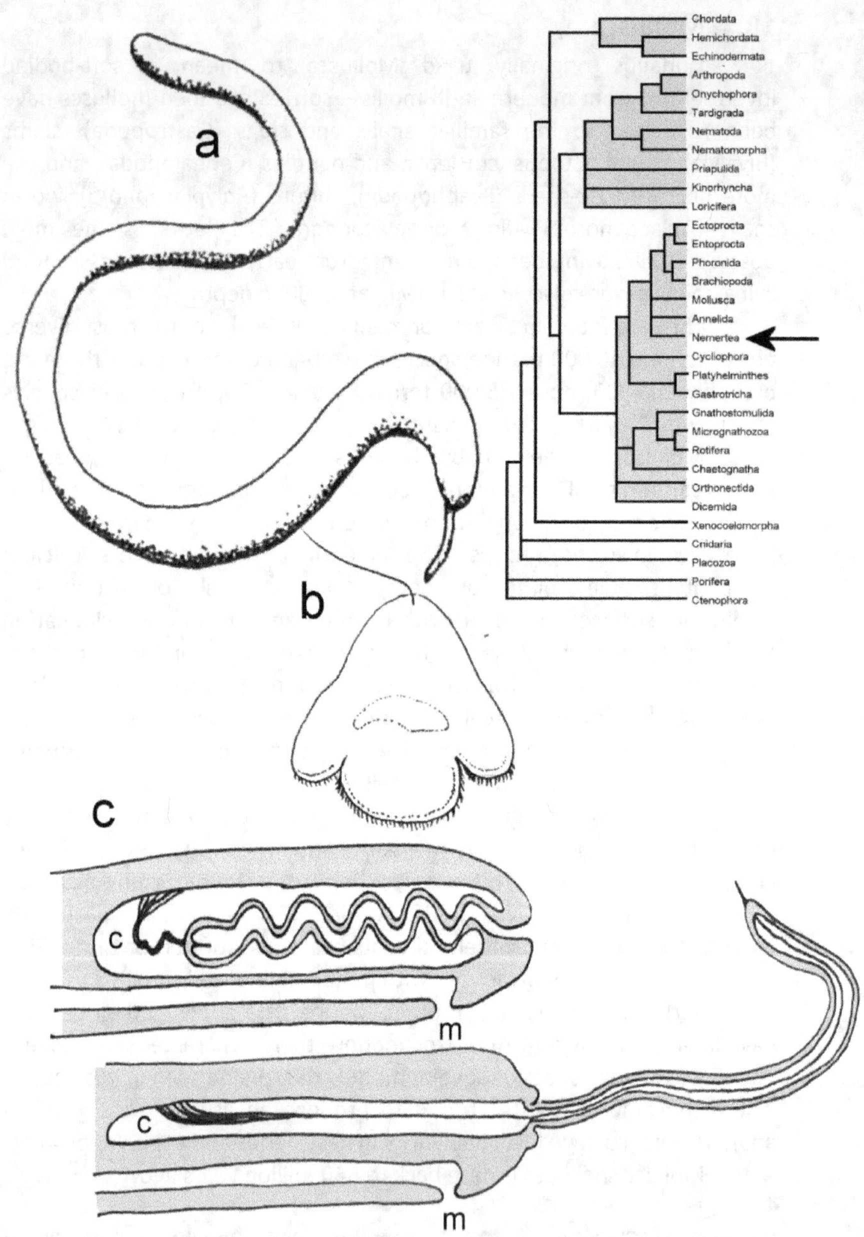

Nemertea: a) adult, b) pilium larva, c) section showing eversion of proboscis

19. Nemertea (ribbon worms)

Nemertea were named more imaginatively than many other phyla; Nemertes was an ancient Greek sea nymph, a link befitting a largely marine phylum, albeit one of largely unattractive worms. Given that Nemertes was reputed to resemble her father Nereus who was sometimes represented as a sort of mer-man with a long eel-like tail, there is some slight connection. Nemerteans are a significant group of 1,300 largely marine predatory worm species, feeding on small invertebrates, or sometimes fish or cephalopods. A handful of species live in freshwater and wet terrestrial habitats. Most are small or microscopic, but they include the longest animal recorded, *Lineus longissimus* at 30 m.

Like acoelomate platyhelminth flatworms they are ciliated and have flame cells for excretion, but they do have a coelom around the eversible proboscis (rhynchocoel) and a closed circulation transporting haemoglobin. One species appears segmented but this is only superficial.

In feeding they use muscular contraction to increase coelom pressure and evert the proboscis rapidly. This stabs the prey and can then be retracted using muscles. Prey are often subdued with toxins. A small number of species are ectoparasites of crabs, living in permanent association with the host and feeding on its eggs.

The brain is limited to a bilobed cerebral ganglion surrounding the proboscis apparatus, rather than the digestive system as is the case in most other animals. Most species have some limited vision, using pigment-cup ocelli but some have rhabdomeres, sometimes with additional lens-like structures, giving them some degree of image forming ability.

They are all sexual, with external fertilisation and direct or indirect development. If a larva is present it is either a short-lived ciliated planuliform larva or the unique pilidium form. 450 species have a pilidum; this helmet-shaped structure effectively acts as a housing for the juvenile. The larval tissues reorganise within it and the fully developed juvenile breaks out from inside this larval casing, in a 'catastrophic metamorphosis'. This larval stage may live for months and feeds on algae, whereas the planuliform version is short-lived and non-feeding.

As they lack any hard parts, the fossil record is very poor. There is a possible nemertean fossil from the Cambrian (505 million years ago).

References: von Dohren & Bartolomaeus 2007; Gittenberger & Schipper 2008

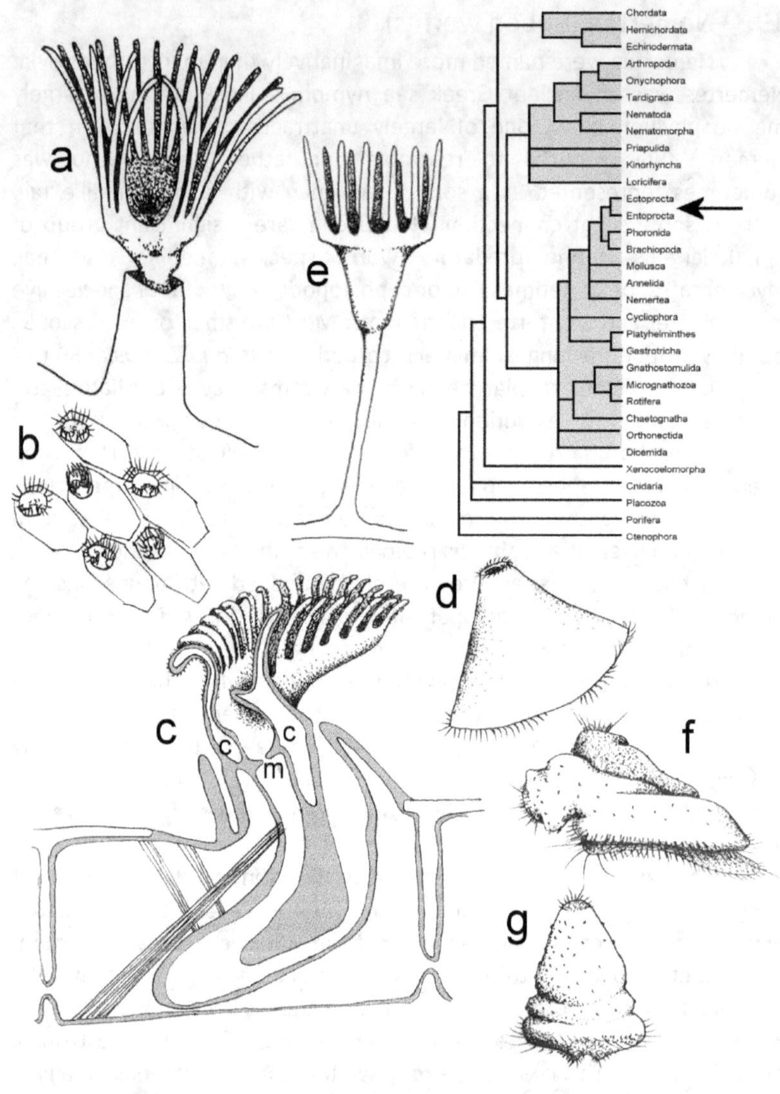

Ectoprocta (a-d), Entoprocta (e-g): b) entoproct colony, c) section showing coelom support to the lophophore, d) ectoproct cyphonautes larva, f) entoproct crawling larva, g) entoproct swimming larva

20. **Ectoprocta + Entoprocta** (moss and goblet animals)

These small filter-feeders are generally similar in appearance. Ectoprocts (or Bryozoa) are colonial animals, usually encased in a calcitic box, whilst entoprocts (or Kamptozoa) are solitary with an exposed cup-shaped body on top of a stalk. Both have a lophophore: a circular or horseshoe-shaped ciliated tentacular crown supported by the coelom. Ectoprocts and entoprocts differ in the former having the anus outside of the lophophore whereas entoprocts have the anus within the tentacular ring. These distinctions give the two phyla their scientific names.

All 6,300 moss animals are benthic, and almost all sessile (a few are creeping species). They are found both in freshwater and marine environments, forming colonies with notable individual polymorphism within the colony. The 18 goblet animal species may be solitary or colonial; all but two are marine, living to a depth of 5,220 m.

Moss animals have a calcitic or chitin skeleton, producing box-like structures encrusting weeds or stones. They lack circulatory, respiratory and excretory structures. There is no distinct brain. Goblet animals do have a pair of protonephridia.

Traditionally moss animals have been placed in the Lophophorata: the phyla with a lophophore (along with brachiopods and phoronids). This has been questioned, although recent studies have revived the possibility that lophohorates may be valid but include the goblet animals as well. There may also be relationships to Cycliophora. Goblet animals were not classed as lophophorates as the lophophore is unusual in not being supported by the coelom (they only have a temporary body cavity), although it is otherwise similar and has the same nervous control.

Moss animals form hermaphrodite colonies although the individual zoids are male or female (but they rarely fertilise within a colony). Fertilisation is external, resulting in a cyphonautes larva. Goblet animals are usually solitary hermaphrodites, and external fertilisation in these results in plaktotrophic swimming larva or a feeding creeping larva.

The first definite moss animals are known from the early Ordovician. They form a major fossil group, with 15,000 species recognised. As they lack a skeleton, the goblet animal fossil record is much weaker, with suggested fossils from the early Cambrian (around 535 million years ago).

References: Laumer et al. 2019; Nielsen 1971, 2012; Reed 1991; Temereva 2017

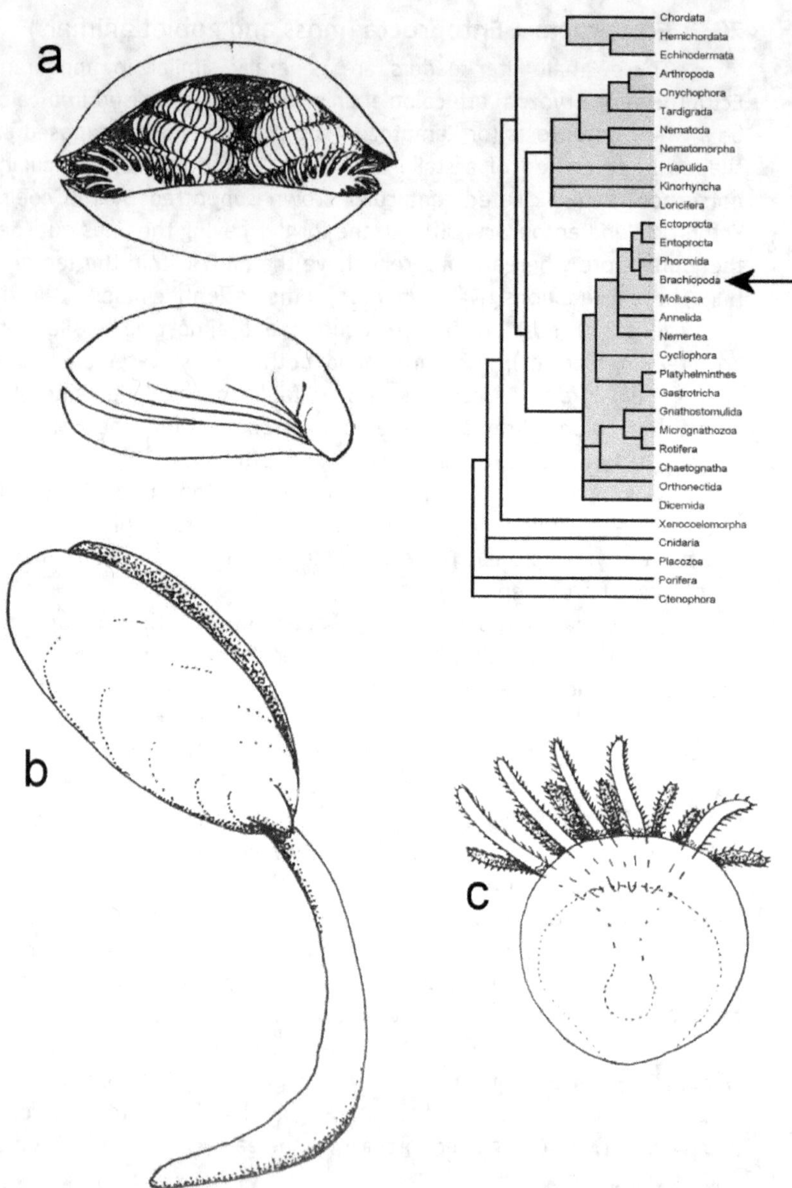

Brachiopoda: a) Articulata, b) Inarticulata, c) larva

21. **Brachiopoda** (lamp shells)

The name Brachiopoda is derived from Greek for arm (brachion) and foot (pous), referring to the tentacles ('arms'). The common name 'lamp shells' derives from a resemblance to Roman oil lamps. Brachiopods are placed in Lophophorata along with ectoprocts and phoronids. They differ in having a bivalved shell, superficially resembling a clam but with dorsal and ventral valves rather than left and right ones like a clam.

Lamp shells are all marine filter-feeders, mostly in cold, deep water. They are very varied in size, from 1 mm to 9 cm (38 centimetres in the fossil genus *Titanaria*). All are sessile, either with a stalk attaching them to the substrate or cemented onto, or into, rock.

The body within the shell has a lophophore and other organs contained within a mantle which secretes the shell, as in molluscs (Chapter 18). The shell is composed of protein, chitin, collagen and layers of calcite or apatite. It is convergent with the mollusc shell, being formed in different ways, although involving some of the same genes.

The lophophore is very variable in shape: horseshoe-shaped, circular or coiled. This filters out plankton and cilia direct the food to a groove leading to the mouth. The gut is normally fully developed although the anus is absent in one group. All gas exchange is carried out through the surface of the lophophore. Oxygen is transported by haemerythrin in the coelom and an open circulation pumped by a heart.

Fertilisation is external (although some females may brood eggs) with gametes being released through the metanephridia which seem not to have a role in excretion, just being used for gamete release. Larvae are either non-feeding or feeding and resemble miniature adults.

Gene expression is extensively modified with Hox gene collinearity being lost in some. Shell formation involves unique genes but also shared ones like chitin synthase (shared with molluscs) and bone morphogen protein (shared with molluscs and vertebrates), indicating a common animal mechanism for mineralisation.

Lamp shells fossilise extremely well and along with molluscs are among the most important fossil phyla 540 million years ago. They were a major group with 12,000 described species but ⅔ were lost in the Permian-Triassic mass extinction. Since then there has been very limited recovery; 443 species are known today.

References: Laumer et al. 2019; Luo et al. 2015; Shimizu et al. 2017; Williams 1997

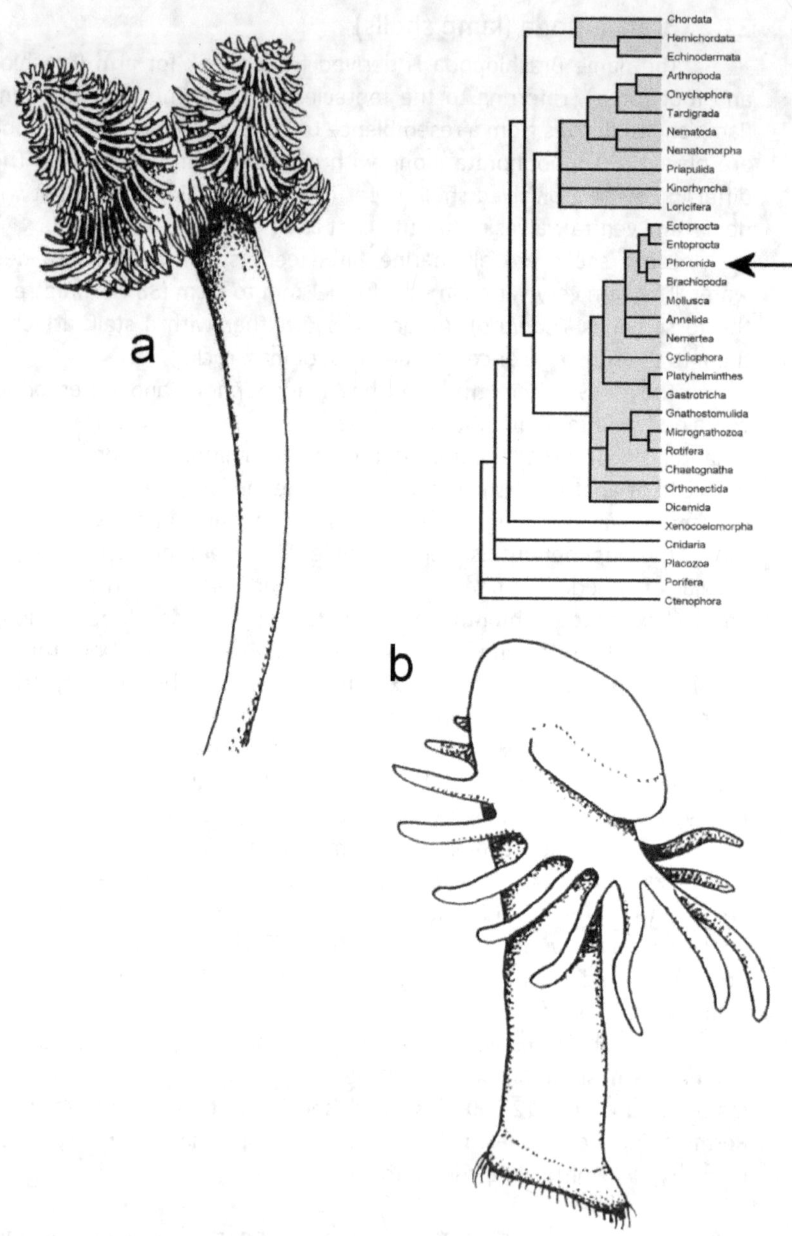

Phoronida: a) adult, b) larva

22.   Phoronida (horseshoe worms)

Phoronida are another phylum of worms named after Ancient Greek mythical beings linked to water. Phoronis is a particularly confused character who seems to be a poorly thought-out compilation of unrelated myths, some of which have nothing to do with the sea. Ovid's highly dubious version takes the tragic Greek nymph Io (or Phoronis), a daughter of a river god, and has her turned into a cow before being reincarnated as a goddess. Bizarrely this ends up being the Egyptian Isis. However confused that was, it was enough for Wright in 1856 to use Phoronis as a suitable watery figure for a phylum of worms.

Phoronids are typical lophophorates, being marine worms with a lophophore. The lophophore is in a characteristic horseshoe-shape, giving the phylum their common name.

Only 15 species have been described even though they are found in all oceans except for the Antarctic, from the shallows to 600 m depth. All are marine, benthic species living in chitinous tubes. They are generally small, from 1.5 to 4.5 cm long, or exceptionally to 50 cm. They are usually solitary, with just one colonial species. This may reach densities of 26,500 per metre.

Structurally they have a flap-like structure, the epistome, which can cover the lophophore, a collar bearing the lophophore and a trunk. The trunk is described as an ampulla, a flask-shaped swelling that serves to anchor the worm in its tube. Like all lophophorates the gut is U-shaped. They are generalised filter feeders, consuming algae, protists and detritus. The circulatory system is closed, with two blood vessels. These are contractile and there is no heart. The oxygen carrier is haemoglobin, enabling them to transport oxygen efficiently in low oxygen environments. Nervous control is achieved with a nerve ring, but there is no distinct brain or notable ganglia.

All are sexual, either hermaphrodite or with separate sexes. Sperm are shed via the metanephridia which also function for excretion. Fertilisation is internal, with egg brooding in one species. The larva is a planktotrophic actinotrocha. One species has a non-feeding slug-like creeping larva and one is viviparous.

Although the worm itself is soft bodied and does not fossilise, tubes resembling those of phoronids date back to the Cambrian around 500 million years ago.

References: Laumer et al. 2019; Temereva 2017

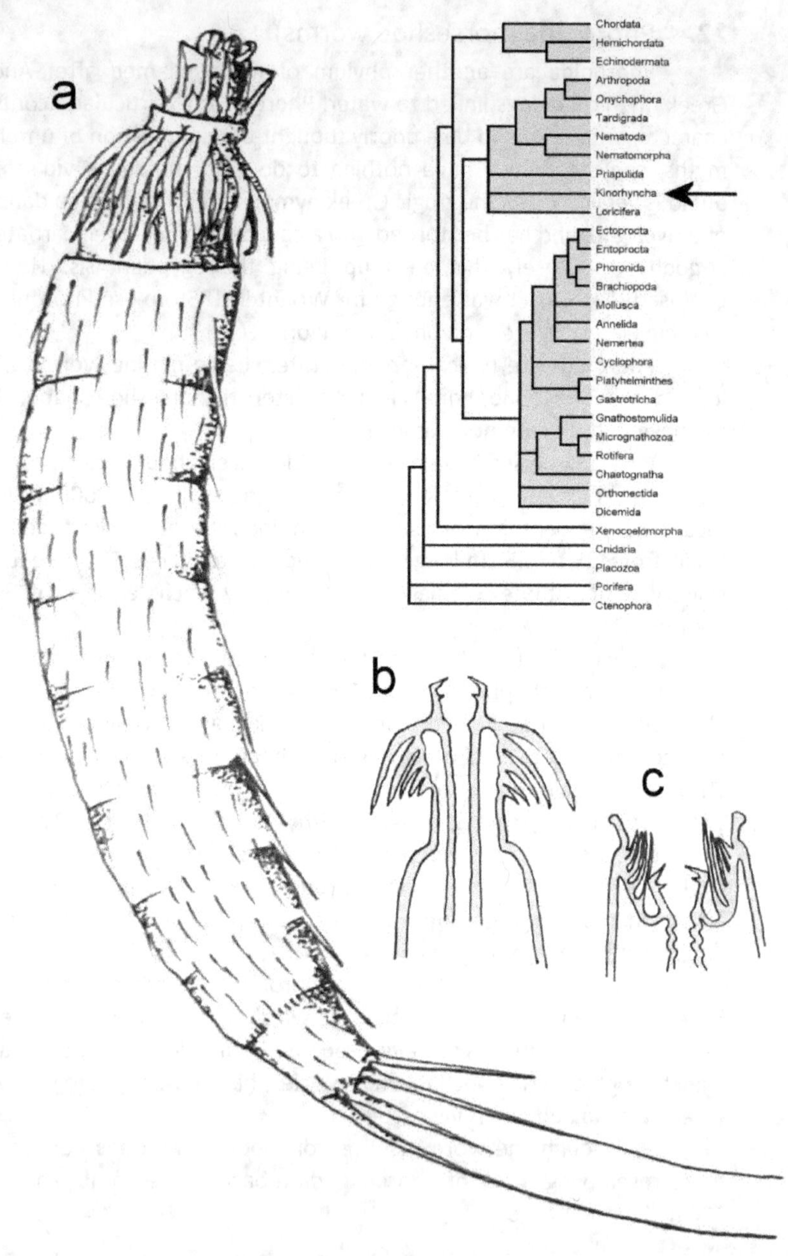

Kinorhyncha: a) animal, b) introvert section extended, c) retracted

23.   Kinorhyncha (mud dragons)

Kinorhynchs are poorly known minute animals found between sand grains in all oceans. They have been found from the intertidal zone down to 7,800 m. 270 species have been described.

They are placed in the Ecdysozoa, a group first proposed in 1997 on the basis of a molecular phylogeny. The most striking feature of the group is cuticle shedding (ecdysis), which gives the group its name. Ecdysozoan cuticle is diagnostic in having three layers: a waxy outer epicuticle and chitin-rich exocuticle and endocuticles. The epicuticle is formed from the tips of the epidermal microvilli. It is moulted under the control of ecdysosteroid hormones in all groups where the control is known.

Around the terminal mouth of most ecdysozoans there is a nerve ring, forming a circumoral brain. This has a similar development and arrangement in all ecdysozoans except in Nematomorpha (Chapter 26) where the gut is dramatically reduced, and in Panarthropoda (Chapter 27). In the latter the brain is ganglionic; the ventral shift in mouth position and anterior concentration of sensory structures (eyes and antennae) are associated with the development of these ganglia.

Although the Ecdysozoa are well supported, their relationships are unclear. Three groups have been proposed: Scalidophora, Nematoida and Panarthropoda. There is little doubt about the last two, but the Scalidophora are more questionable. This comprises the Kinorhynchia, Priapulida and the minute, poorly known Loricifera. All have a protrusible or telescopic spiny introvert feeding structure, the spines (or scalids) of which give the group its name. The problem with this group centres on the question of whether the Loricifera really belong here (Chapter 24).

Mud dragon bodies are in the form of a feeding introvert, a neck region and 11 trunk segments. The spiny introvert can be protruded to catch prey. As minute animals there is no need for a circulatory system beyond a pseudocoelom, and some are completely acoelomate.

Fertilisation is internal, but no details are known. Development is direct, with no larval stage.

There are suggested fossil remains from 535 million years ago, but due to their very small size their identity is uncertain.

References: Adrianov & Maiorava 2015; Aguinaldo et al. 1997; Brown 1983; Schmidt-Rhaesa et al. 1998; Zhang et al. 2015

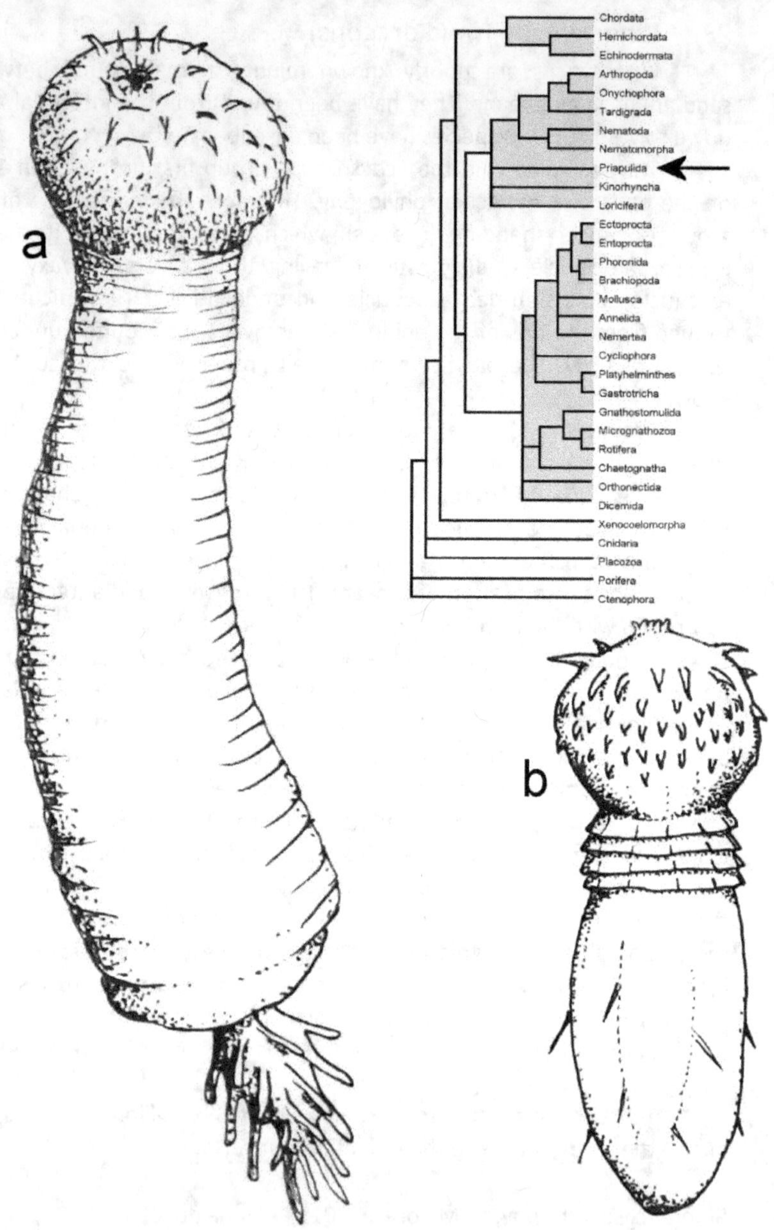

Priapulida: a) adult, b) loricate larva

24. Priapulida (penis worms)

Priapulids are named after the Ancient Greek male fertility god Priapus, for obvious reasons. In 1758 Linnaeus named the first species *Priapus equinus* and *P. humanus*. However, the former is a sea anemone and the latter included sea cucumber specimens. Accordingly, in 1816 Lamarck renamed it *Priapulus caudatus*, the type species of Priapulida.

Priapulids are burrowing marine worms, varying in size from 1 mm to 40 cm in length. They are commonly intertidal but have been found down to 5,700 m depth.

22 species have been identified. They feed on detritus and small invertebrates that come within reach of the proboscis. They can reach high densities; 85 individuals per metre have been recorded.

The worm-shaped body is divided into a cylindrical trunk and a telescopic spiny proboscis with cuticular teeth that is used to grasp prey. The body has an annulated alpha-chitin rich cuticle which is periodically shed. The brain is a typically ecdysozoan ring around the pharynx, with unpaired ventral longitudinal nerve cords. The muscles are circular and longitudinal. The circular muscles are spaced out, causing an annulated appearance of the ectoderm, although penis worms are unsegmented. They lack a distinct circulation but have a large pseudocoelom containing amoebocytes and erythrocytes with haemerythrin, which provides oxygen transport.

Penis worms are all sexual, with separate sexes. The gametes are released through flame cells and fertilisation is external. Development in this phylum is unusual in that the anus develops from the blastopore as in the deuterostomes (Chapter 30). In one species development is direct, but in all the others there is a loricate larva. This has a spiny head structure, a pleated collar and a vertically grooved trunk with a few large setae and bears a striking resemblance to the Loricifera phylum (Chapter 24). However, the details of the armature of the pharynx differ.

Despite being soft bodied, remarkably well preserved fossils of the Burgess Shale 508 million years ago are convincing priapulids. In addition fossils of what seem to be their cuticular spines and their burrows date from the Cambrian 535 million years ago. The largest priapulid (at 2 m) is the fossil species *Omnidens amplus*.

References: Hou et al. 2006; Martín-Durán et al. 2012

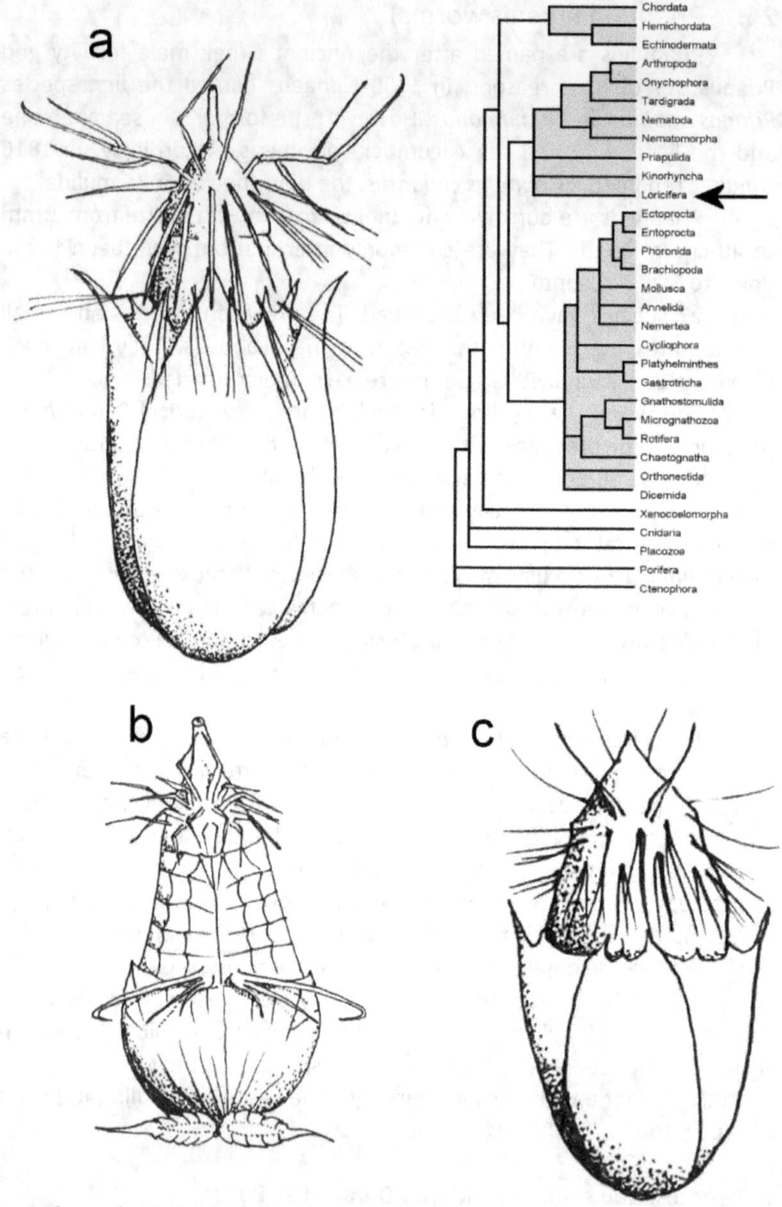

Loricifera: a) adult, b) Higgins larva, c) postlarva

25. Loricifera (jewel animals)

Loriciferans are named for the cuticular plates which resemble a 'lorica' or armoured cuirass, specifically the transverse metal strips worn by Roman legionaries.

Loriciferans are microscopic predators living between sand grains on the sea bed. All 38 species are under 0.8 mm in length. They were first found in 1974 as larvae, but the phylum was not described until 1983 when adults were found. They are mostly recorded from coarse sediment 20-450 m below the surface but one has been found at 8,000 m depth. They are predators of bacteria and algae. The body is composed of a mouth cone, head, neck, thorax and abdomen with a protective cuticular casing of six plates (the lorica). The mouth cone is the feeding apparatus; a protrusible spiny introvert. These are the only animals lacking mitochondria. Like many protists that have lost mitochondria they have hydrogenosome-like derivatives of mitochondria. Not relying on oxygen-based respiration they can live in completely anoxic sediments.

They usually have separate sexes and internal fertilisation. The larva of most species is called a 'Higgins' larva after Robert Higgins who discovered the first specimen in 1974. This has caudal locomotory 'toes' giving it a unique movement pattern. The larva moults several times into a 'postlarva', then into the male or female adult.

Loriciferans resemble priapulid larvae (Chapter 23) and have been suggested to be paedomorphic (breeding in juveniles form) priapulids. Under this idea larval priapulids would become sexually mature without metamorphosing into the normal adult as a result of an accelerated gonad development (the progenetic version of paedomorphosis, as opposed to neoteny where somatic development is retarded). However, it would be odd for a paedomorphic larva to evolve its own larva. Furthermore, as with the suggested paedomorphic origin of rotifers (Chapter 12), the proposal is based purely on a superficial similarity. Similar untested claims have been made for the origins of flatworms, nematodes, molluscs and even insects. The idea of a phylum arising from paedomorphosis has been tempting as, in theory, a small change in a developmental regulatory gene could cause what appears to be a major morphological change extremely rapidly.

There is a fossil proposed to be a 490 million years old loriciferan.

References: Danovaro et al. 2016; Kristensen 1983, 2002; Neves et al. 2016; Warwick 2000

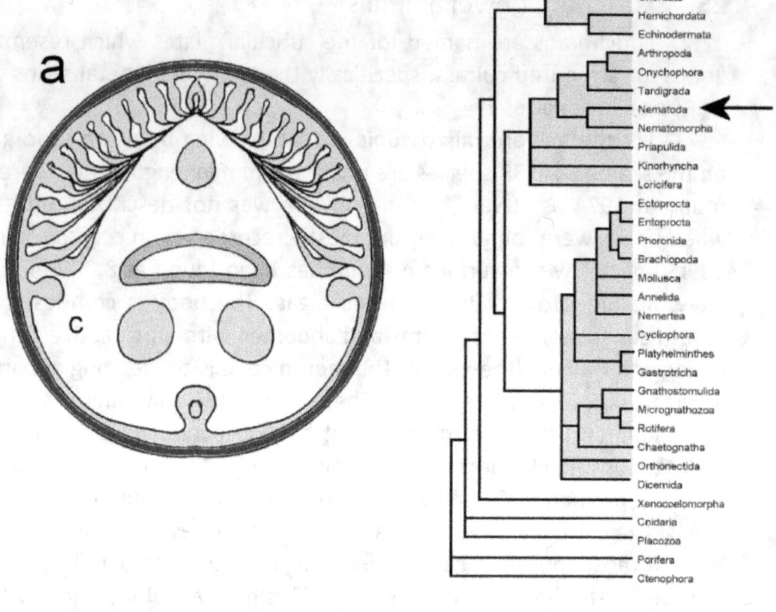

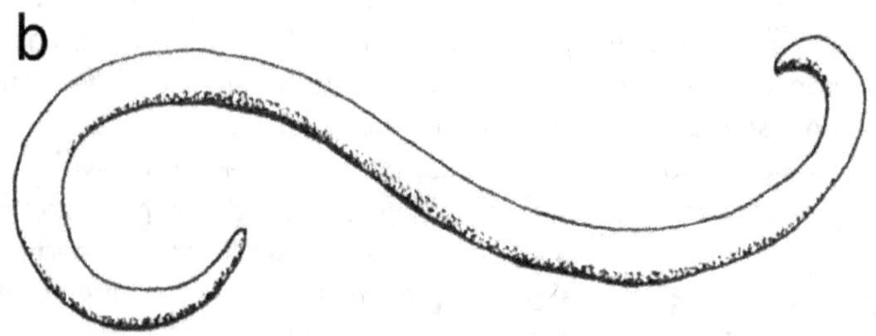

Nematoda: a) section (see page 4), b) animal

26. Nematoda (round worms)

Nematodes are the round worms or thread worms (nema being Greek for thread) with many species of agricultural and human health significance, along with the genetics model animal *Caenorhabditis elegans*. They are the sister group to the very similar Nematomorpha (Chapter 27), the two phyla being grouped as the 'Nematoida'. As with all ecdysozoans, there is a cuticle, but in these two phyla its chitin component is restricted to the pharynx. It is supported by layers of crossing collagen fibrils. They are also usual in ecdysozoans in lacking circular muscle in the body wall.

Round worms may be free-living or parasitic, found in all ecosystems. They are highly abundant, making up an estimated ¾ of all animal individuals. They are also exceptionally diverse: 26,650 species have been described but it is speculated that there may several million more. Most are very small (the smallest just 82 microns) but parasitic species can be large, up to 9 m in *Placentanema gigantissma*, a parasite of the placenta of the sperm whale *Physeter macrocephalus*.

They are structurally simple and unsegmented, although the cuticle is annulated in some free-living forms. The cuticle is shed under the influence of ecdysone three or four times in their lives. There is no specialised circulatory system, relying on the large, fully-open pseudocoelom for transport. Some small species are acoelomate.

Given the round worm shape and lack of circular muscle, they are constrained to living in constricted spaces (e.g. in the soil, between cells or in animal guts and blood vessels). Their longitudinal muscles work antagonistically against the environment and the pressure of the pseudocoelom. This has an extraordinarily high pressure, usually above 65 mmHg. This pressure is not known outside of birds and mammals, other animals being in the range of 1-21 mmHg. This body pressure requires other adaptations, including a remarkable one: rather than their sperm being swimming with flagellae they crawl in an amoeboid manner!

Sexual reproduction is varied in round worms; they may have males and females; males and hermaphrodites; males, females and hermaphrodites; or just hermaphrodites. Some are exclusively asexual.

Fossils of parasitic forms have been found in amber dating from the early Cretaceous, the oldest possible nematode is early Devonian.

References: Bottino et al. 2002; Decraemer et al. 2014; Harris & Crofton 1957; Hermosilla et al 2015; Moens et al. 2014

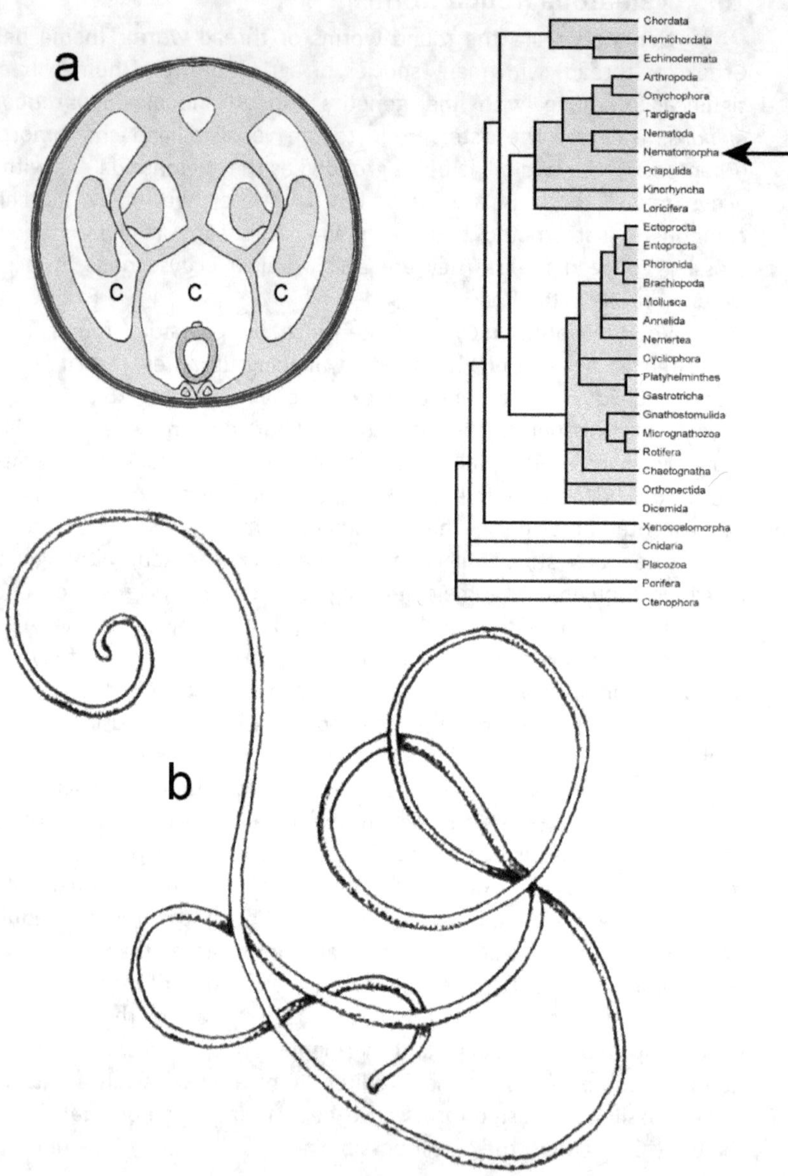

Nematomorpha: a) section (see page 4), b) animal

27. **Nematomorpha** (Gordian worms)

As with Nematodes, Nematomorphs belong in the Nematoida, sharing similar cuticles and musculature. They differ in this being an entirely parasitic group. 355 species are known, 5 parasitise marine crabs and lobsters, the others alternate between freshwater and terrestrial arthropods.

Adults mate in a mass in water, the knotted ball leading to the common name of Gordian worms. The alternative name 'horsehair worms' refers to their appearance, being very long and extremely thin (50-100 mm long, only 1-2 mm wide). Similarly, the scientific name translates to 'thread-shaped worms'.

As specialised parasites the gut is reduced, in some cases the mouth may be lost. There is no distinct circulation or excretory organs. The cuticle is moulted only once, when changing from the juvenile to the thicker adult form.

Reproduction may be asexual, or sexual with hermaphroditism. The marine species are capable of swimming using setae and they seek out crabs and lobsters. In freshwater species adults are free-living, producing eggs which develop into free-swimming gordioid larvae. These are swallowed by aquatic insects, snails or fish, pass through the gut wall and encyst in the gut tissue. This intermediate host is ultimately consumed by a terrestrial arthropod, either through the host moving onto land as part of its life cycle (e.g. an insect larva) or as a result of being semi-aquatic (e.g. snails). The main definitive hosts are crickets, cockroaches, mantises and beetles. In the definitive host the cyst develops into the non-feeding adult form which eventually exits the host and moves into water to breed.

Both Gordian worms and nematodes have the a reduced complement of Hox genes, just five in Gordian worms and a maximum of six in nematodes (reduced to four in some species) scattered over one chromosome. Nematodes also reduce ParaHox to a single gene. The simplified structures of these two phyla makes several of the Hox genes largely redundant (effectively neutral), and so vulnerable to loss by genetic drift.

The fossil record of early nematomorphs is disputed, with the exception of amber preserved fossils, which date from the Cretaceous 110 million years ago.

<u>References</u>: Hanelt et al. 2005; Poinar & Buckley 2006; Yoshida et al. 2017

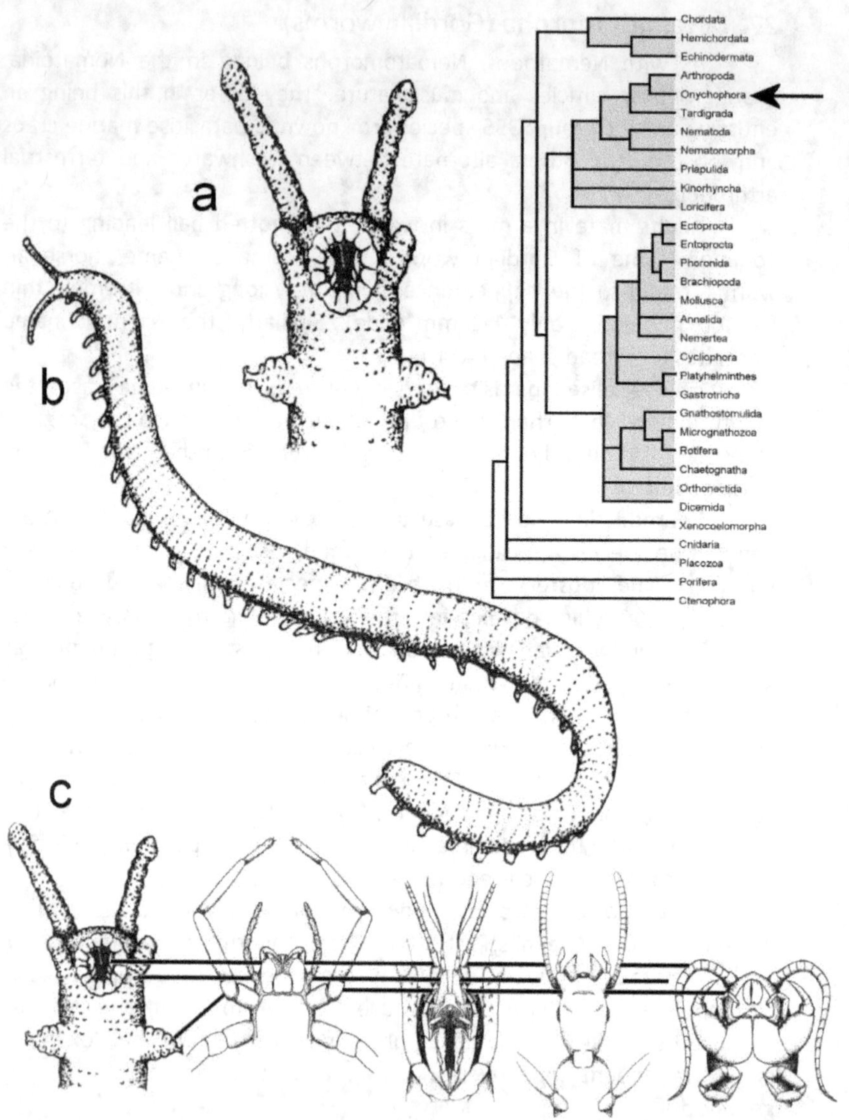

Onychophora: a) head detail, b) animal, c) homology of panarthropod head appendages indicated by Hox genes: jaw – chelicera – antenna 1, slime papilla – palps – antenna 2 – (absent), leg 1 – leg 1 – maxilla, in (left to right) Onychophora, Chelicerata, Crustacea, Insecta, Myriapoda

28. Onychophora (velvet worms)

Onychophora are named after their claws: the 'claw bearers' (onux + phoros). Onychophorans are the terrestrial predatory velvet worms, with a worm-shaped body which appears unsegmented externally. Pairs of claw-bearing fleshy legs run along the body, corresponding to the internal segments.

They are grouped with tardigrades (Chapter 27) and arthropods (Chapter 28) as the Panarthropoda. All have a ganglionar brain above the oesophagus. In onychophorans this has two segments: protocerebrum and deutocerebrum. The protocerebrum innervates the antennae, the simple rhabdomeric eyes and jaws. The jaws are the appendages of the second body segment, equivalent to either the jaws (in chelicerates) or first antennae of arthropods. The antennae of their first segment correspond to the feeding structure of tardigrades.

Velvet worms are predators of small invertebrates. Prey are captured using a glue-like slime that is ejected by slime-papillae on the head. The immobilised prey are punctured using the chitinous jaws.

The coelom is reduced to a space around the gonads and nephridia. An open circulation is present as a haemocoel, the dorsal part of which (the paricaridal sinus) contains a muscular vessel functioning as a heart. They use haemocyanin for oxygen transport. Respiratory structures are in the form of tracheae which lack spiracles.

Velvet worms have the full set of protostome Hox genes and seems to have an arrangement very close to the ancestral panarthropod. It appears that the true arthropods have modified their Hox gene control resulting in greater separation of anterior and posterior regions (the more distinct head/body division of arthropods) and more variation between segments.

Reproduction in some species is viviparous, with a placenta.

180 species have been described, they are found throughout the tropics and subtropics. They are the only exclusively terrestrial phylum, but there are marine fossils from 520 million years ago. Terrestrial fossils date from 310 million years ago.

References: Garwood et al. 2016; Janssen et al. 2014 ; Mayer et al. 2010, 2015

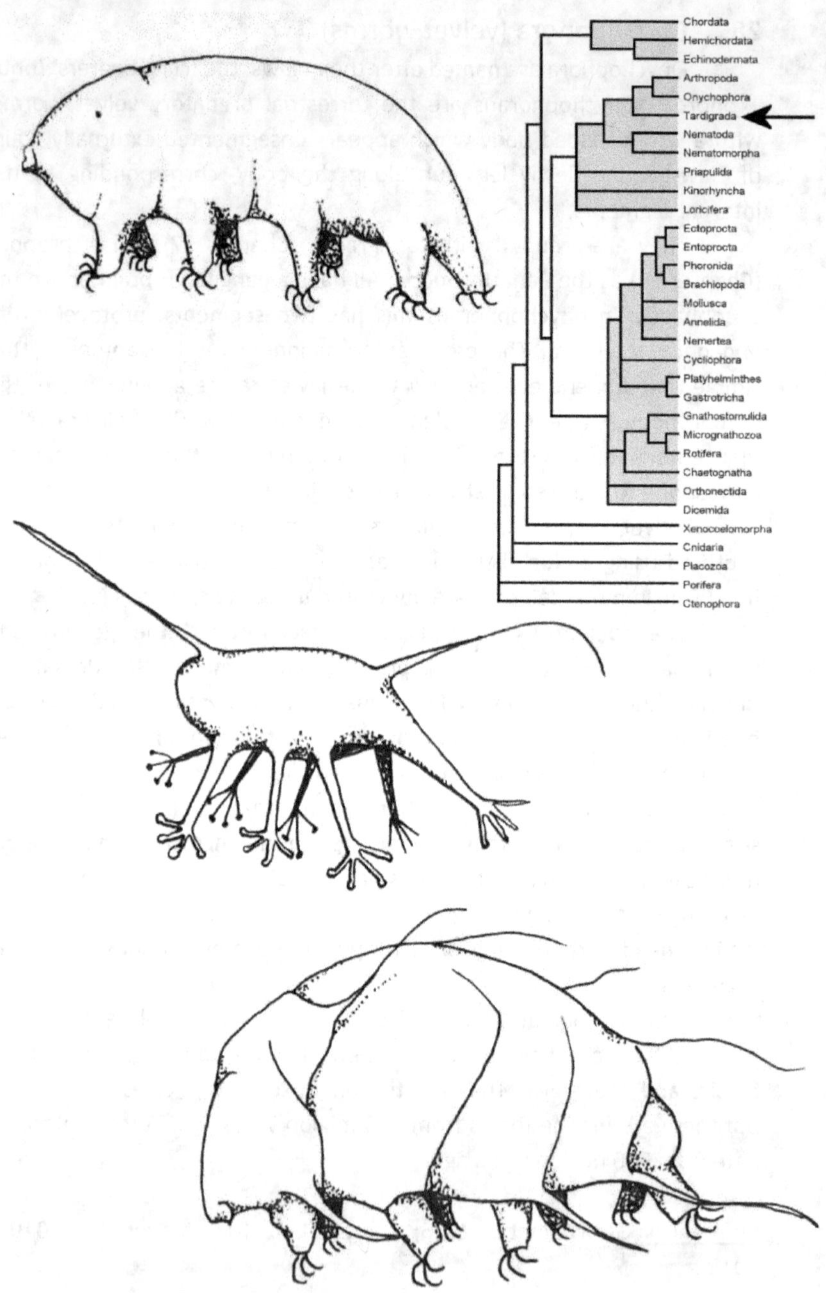

Chordata
Hemichordata
Echinodermata
Arthropoda
Onychophora
Tardigrada
Nematoda
Nematomorpha
Priapulida
Kinorhyncha
Loricifera
Ectoprocta
Entoprocta
Phoronida
Brachiopoda
Mollusca
Annelida
Nemertea
Cycliophora
Platyhelminthes
Gastrotricha
Gnathostomulida
Micrognathozoa
Rotifera
Chaetognatha
Orthonectida
Dicemida
Xenocoelomorpha
Cnidaria
Placozoa
Porifera
Ctenophora

29. Tardigrada (water bears)

Tardigrada are named for their characteristic movement, a 'slow walk' (Latin tardus + gradi). Tardigrades are usually regarded as panarthropods along with onychophorans and arthropods, although some schemes group them with nematodes. The difficulties here may be due to extensive genome reorganisation and long-branch attraction.

1,300 species have been described, but they are poorly studied in many areas and ecosystems. Most species are terrestrial, where they occur in leaf litter and moss. They are also found in aquatic systems.

All are under 1 mm and this miniaturisation has arisen though loss of the mid-body region, their segments corresponding to the 5 of the arthropod head and the terminal one. Their Hox genes have been correspondingly reduced. Similarly, the brain resembles that of other panarthropods but is restricted to a single segment. Being so small they have no need for circulation beyond a pseudocoelom.

Water bears lack antennae and cuticular jaws, having a unique pumping pharynx containing a stylet instead. Based on Hox gene expression their first pair of legs is homologous to the jaws of onychophorans and the jaws or first antennae of different groups of arthropods. Although antennae are absent they do have eyes, in the form of pigmented cup-shaped structures.

Water bears are the hardiest of all animals, being tolerant of extremes of temperature, desiccation, radiation and even placement in a vacuum. Dormant tardigrades were sent into space; on return to Earth most were successfully revived. Their resistance arises from anhydrobiosis where they produce a dormant 'tun' which has remove almost all body water, reducing volume by 87%. This is achieved by various biochemical adaptations, including accumulation of osmolytes (e.g. trehalose) to protect against dehydration damage, and intracuticular lipids as a permeability barrier. This form can survive for 30 years.

Reproduction is sexual, usually with separate sexes. Fertilisation is strangely both external and internal. The female lays eggs between her body and an old outer layer of cuticle. The male then fertilises the eggs under the old cuticle, technically outside of the female's body, but still within her cuticle. Development is direct.

There may be tardigrade microfossils from 520 million years ago.

References: Gross & Mayer 2015; Halberg et al. 2013; Hashimoto et al. 2016; Smith et al. 2016

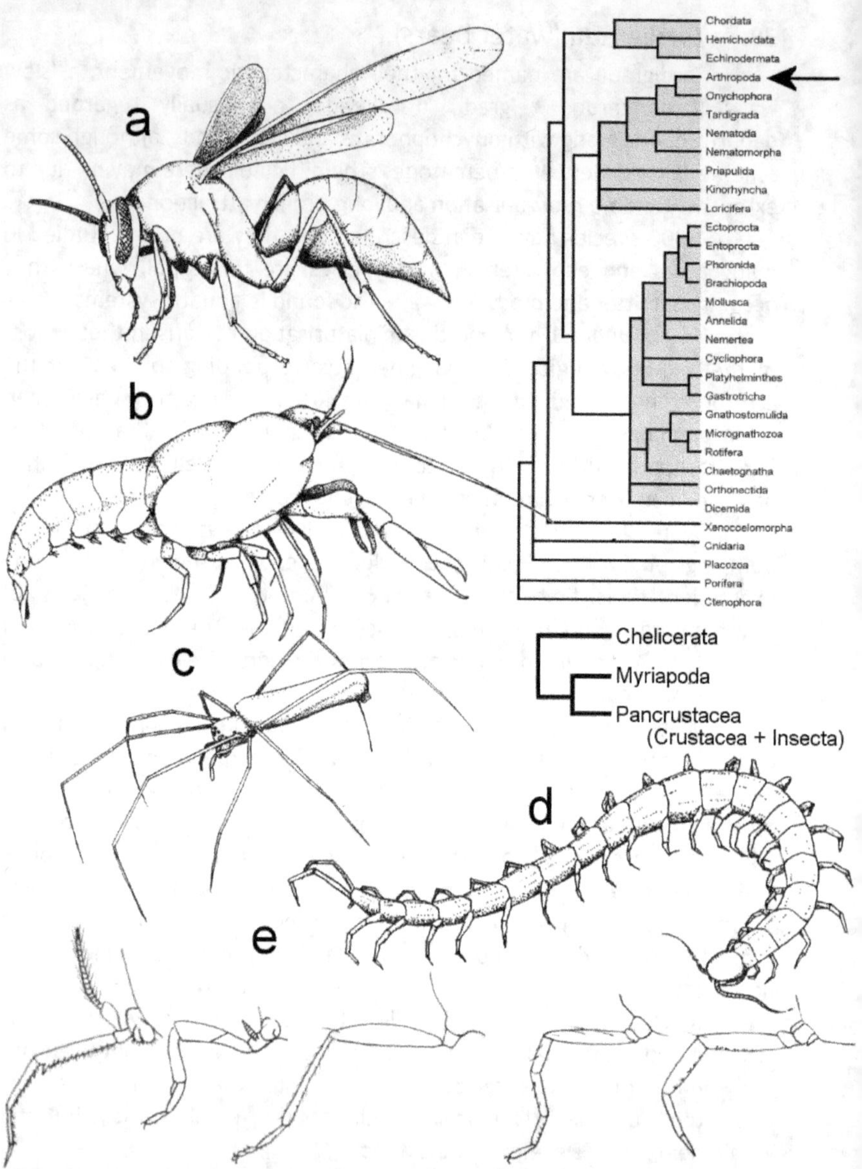

Arthropoda: a) Insecta, b) Crustacea, c) Chelicerata, d) Myriapoda, e) biramous and uniramous legs in (left to right) Crustacea, archaeognathan insects, Insecta, Chelicerata, Myriapoda

30. **Arthropoda** (arthropods)

Arthropods are probably the most familiar invertebrate phylum, comprising the four subphyla of insects, crustaceans, myriapods (millipedes, centipedes, pauropods and symphylans) and chelicerates (sea spiders, horseshoe crabs and arachnids). The most diverse group is the insects, with over one million described species, but true diversity of arthropods is likely to be between 5 and 10 million species. The arthropod subphyla were traditionally grouped based on jaw and leg structures but these conflict with most molecular studies. These now indicate that they should be divided into the chelicerates and mandibulates. Within the latter, insects lie within the crustaceans, the two groups together forming the Pancrustacea.

Arthropods are grouped with tardigrades and onychophorans as the Panarthropoda, sharing segmentation and a similar brain. This is divided into three segments: the protocerebrum, deutocerebrum and tritocerebrum. The protocerebrum innervates the antennae, mouthparts and compound eyes. These are modified to simple eyes in arachnids and myriapods. The segments are grouped into distinct regions, tagma, which vary in number from three (head, thorax, abdomen) in insects to two in chelicerates. The cuticle is chitinous and tanned and mineralised variously in different groups, all are moulted under control of ecdysone.

Circulation is well developed in all groups, with the coelom restricted to an open circulation of haemolymph pumped by a heart. Gas exchange is achieved using various structures: gills, gill books, book lungs or tracheae. Oxygen is transported by haemocyanin.

All arthropods have jointed limbs, these may be simple structures (uniramous) or branched (biramous). In the latter the leg base branches into a main walking or swimming limb and an accessory gill or gill cleaning structure. Living taxa are all supposed to be uniramous except for the crustaceans. Given that the insects are now known to be descended from biramous ancestors (e.g. archaegonathans) uniramy is clearly derived. Developmental genetics show that this is not from the loss of a branch of the leg, but from suppression of the subdivision that leads to biramy.

All arthropods have internal fertilisation. Crustaceans have free-living larvae. Insect larvae have evolved as a secondary feeding stage.

The exoskeleton means that fossils are abundant from 521 million years ago.

<u>References</u>: Edgecombe & Legg 2014; Stork 2018

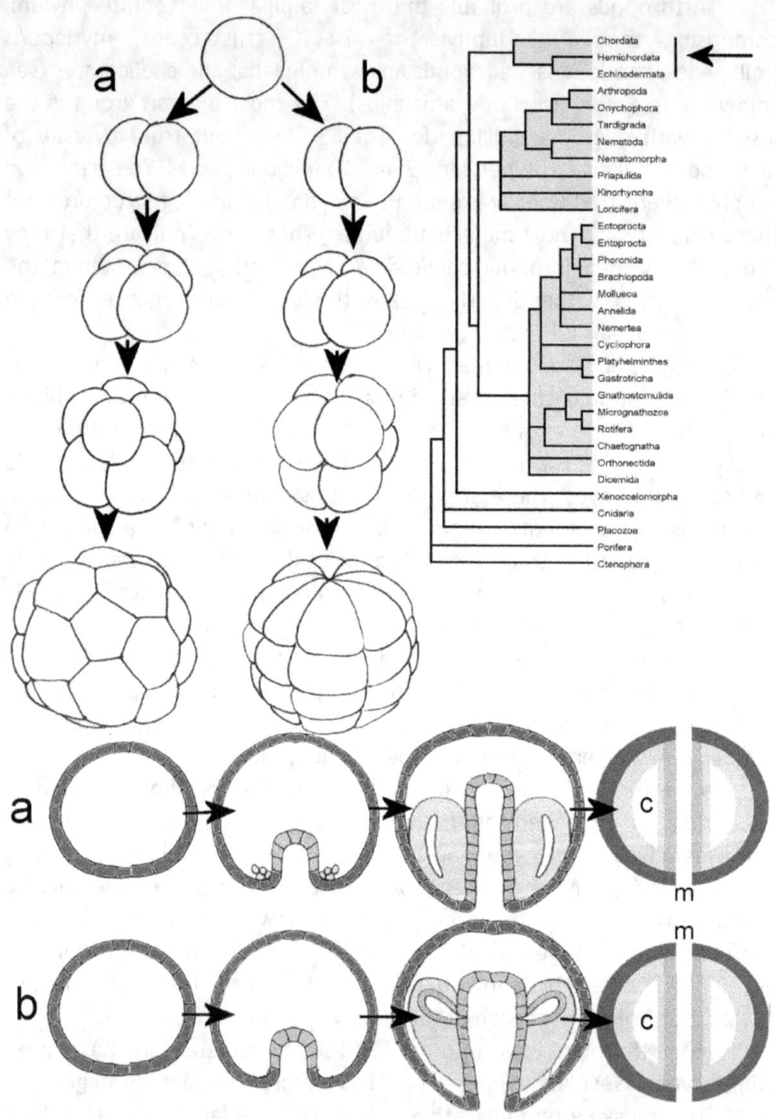

Summary of theoretical embryonic development in protostomes (a) and deuterostomes (b) (key to sections – see page 4)

31.   **Deuterostomia** – embryology and groups

The deuterostomes comprise three living phyla: Echinodermata, Hemichordata and Chordata. Several Cambrian fossil groups have been suggested to be deuterostomes but all are questionable: vetulicolians might be arthropods; yunnanozoans might be kinorhynchs; cambroernids and vetulocystids may be deuterostomes, but are difficult to interpret.

All deuterostomes are triploblastic coelomates. Symmetry varies from radial (echinoderms) to bilateral (hemichordates and chordates). The through-gut is fully developed except in a few species.

The group is defined by its embryonic development. The deuterostome condition (meaning 'second mouth') of the mouth being the second gut opening to develop is better viewed as the anus forming from the original blastopore. The mouth is a new structure developing at the base of the embryonic gut. The cleavage pattern is radial. Mesoderm forms from cells migrating from the endoderm with the body cavity developing within these pockets from the archenteron (enterocoely). The development is not reliable as the mouth formation pattern is also found in some protostomes, radial cleavage is present in ctenophores (Chapter 5) and some protostomes. Hemichordates (Chapter 32) have been reported to develop the body cavity from mesodermal splits (schizocoely).

One of the few anatomical synapomorphies is the presence of pharyngeal gill slits. These are present in hemichordates and at least in embryonic chordates. They are probably lost secondarily in echinoderms (Chapter 31) as early echinoderm fossils appear to have gill slits. This is further supported by the gill slits being formed by the action of a cluster of four transcription factor genes, which are found in all deuterostomes. In hemichordates the gill slits are used in filter-feeding.

The earliest deuterostome may be the strange creature *Saccorhytus coronaries*. This 540 million year old fossil is of a globular animal just 1.3 mm in diameter. It appears to have been a filter-feeder with a large central mouth in a ring of eight small pores (possible gill slits). No anus seems to have been present. It is thought to be a deuterostome based on superficial similarity to the fossil vetulicolians and vetulocystids. If reconstructions of these two groups as deuterostomes are correct, they also lacked a through-gut and possessed gill slits. Reliable deuterostome fossils date from 540 million years ago (although Ediacaran fossils have been suggested to be deuterostome from 600 million years ago).

References: Dilly 2014; Han et al. 2017; Nielsen 2012; Simakov et al. 2015

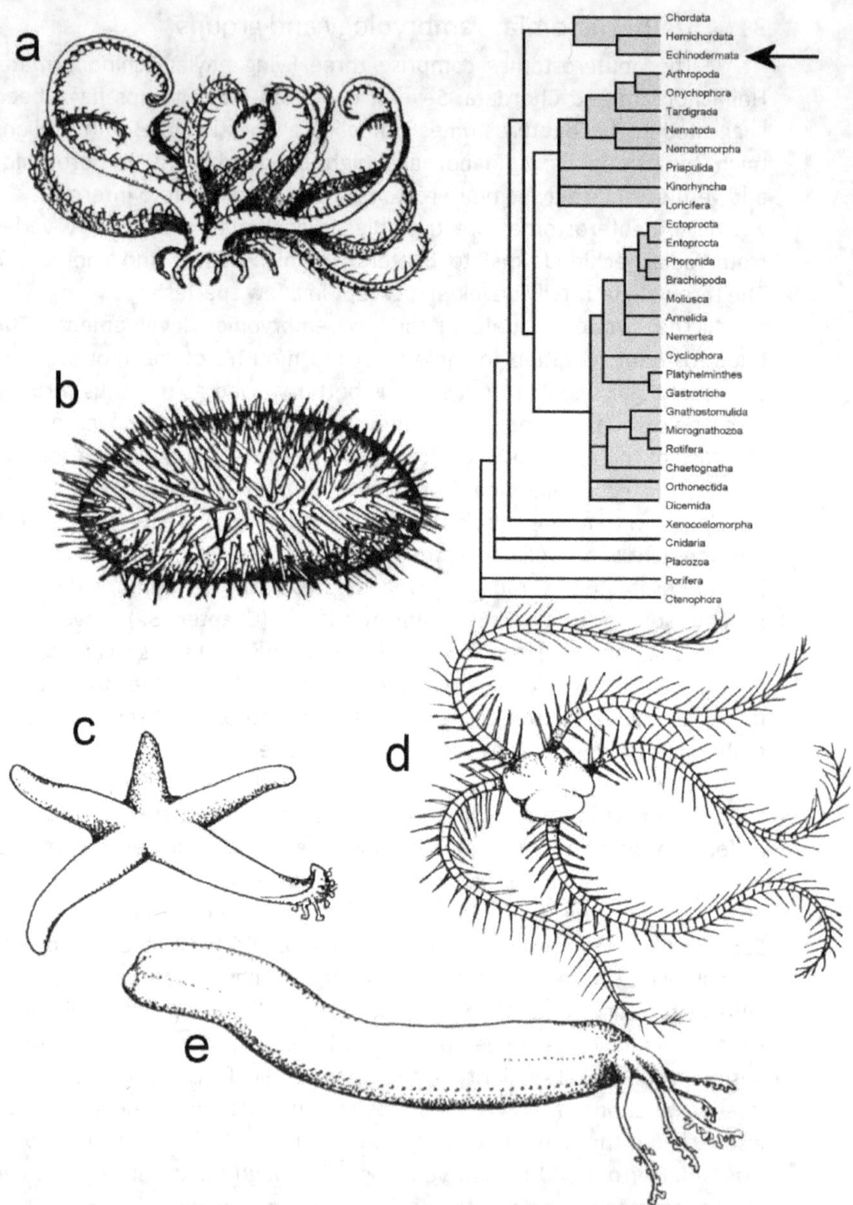

Echinodermata: a) Crinoidea, b) Echinoida, c) Asteroida, d) Ophiuroida, e) Holothuroidea

32. **Echinodermata** (echinoderms)

Echinoderms are an entirely marine phylum comprising the starfish, brittle stars, sea lilies, sea cucumbers and sea urchins. The latter give the group their name, from the Greek ekhinos (hedgehog or sea urchin), combined with derma (skin). The 7,300 species are carnivores (starfish), herbivores (sea urchins), detritivores (sea urchins, sea cucumbers) or filter-feeders (brittle stars, sea lilies).

Structurally they are varied, being spherical (sea urchins), with radiating arms (starfish, brittle stars, sea lilies) or worm-shaped (sea cucumbers). They are thick skinned, with a body wall containing collagen that in sea-cucumbers can be relaxed or stiffened under nerve action, enabling them to squeeze into tight spaces. This 'catch connective tissue' is more localised in other groups, these are less flexible, with a mesodermal calcium carbonate derived skeleton. Locomotion is effected by the tube feet that connect to the water vascular system, a modified coelom. Tube feet also excrete ammonia, as does the respiratory tree of sea cucumbers (tubes circulating water within the cloaca for gas exchange). There is no other osmoregulatory mechanism and the lack of osmoregulatory flexibility prevents them from adapting to freshwater.

The gut is normally well developed but is a bottle-gut in some starfish and all brittle stars. Many predatory starfish evert their stomachs over their prey and digest extracellularly. Herbivorous sea urchins have complex calcitic jaw plates and rods (the Aristotle's lantern) for grazing.

Light sensing is possible in some brittle stars, sea urchins and starfish using light-sensitive cells and calcitic 'micro-lenses'. These are scattered over brittle stars, or restricted to the feet of sea urchins and the tips of starfish arms. Starfish can have up to 200 lenses gathered into ommatidia and low resolution image formation is possible.

Echinoderms are unusual in being radial, generally pentaradial. Developmentally they start as bilateral larvae, developing radiality later. Many echinoderms have extensively reordered their Hox genes, this is not associated with radiality as some radial echinoderms retain collinearity.

Most species have separate sexes and external fertilisation. Development is typically deuterostome, producing very diverse larvae.

Due to their calcium skeleton the fossil record is extensive, with 15,000 species from the past 520 million years. Early forms were bilateral or asymmetrical, with pharyngeal gill slit feeding systems.

<u>References</u>: Garm & Nilsson 2014; Schiemann et al. 2015

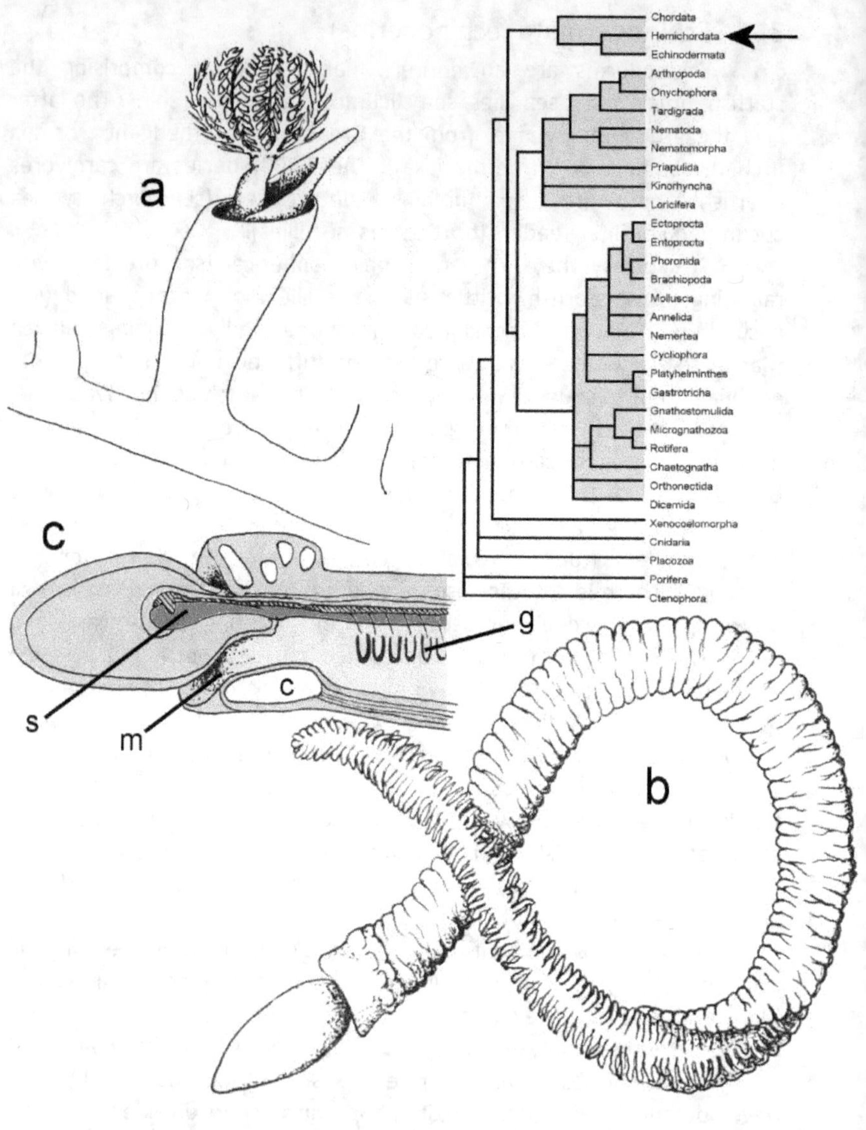

Hemichordata: a) Pterobranchia, b) Enteropneusta, c) section through anterior of an enteropneust (c – coelom, g – gill slits, m – mouth, s – stomochord, dark shading)

33.   Hemichordata (acorn worms & sea angels)

Hemichordata are the only phylum to be named in reference to a different phylum, hemichordate literally meaning 'half a chordate'. This is often taken to refers to the short stomochord skeletal structure that resembles the notochord of the Chordata (Chapter 34). In reality Bateson proposed the name (as a subgroup of Chordata) in 1885 because he thought all the key features resembled those of a chordate with "partial or arrested development". These similarities are a mixture of superficial resemblances, ancestral characters and convergences.

Hemichordates comprise two very different groups: acorn worms (Enteropneusta) and sea angels (Pterobranchia). Both are exclusively marine, acorn worms feeding on detritus in sediment and sea angels as sessile colonial filter-feeders. There are 140 species, all in shallow waters. Acorn worms live in U-shaped burrows in the mud, using cilia to direct organic debris to the mouth; the collar sorts edible material from detritus. The body has three regions: proboscis, collar and trunk, with a separate coelomic compartment in each. In acorn worms the proboscis forms the 'acorn' of their name and in sea angels it is the filter-feeding tentacle set.

In acorn worms the one coelomic cavity develops by splitting of the mesoderm (schizocoely) although the others develop in normal deuterostome off-pocketing from the gut (enterocoely). All sea angel coeloms develop by schizocoely. The proboscis coelomic pocket forms part of a unique excretory system, the 'heart-glomerulus complex'.

Both groups have pharyngeal gill slits, as is ancestral to all deuterostomes. The circulation is well developed but open, as befits a low metabolism life-style. The nervous system takes the form of a ventral and a short dorsal collar nerve cord. A brain forms in the ectoderm of sea angels but not in acorn worms where no distinct brain develops but the dorsal nerve cord is internalised in the collar and resembles the chordate neural tube (Chapter 32).

Reproduction can be asexual or sexual. They are hermaphrodite, with external fertilisation. Development is indirect with tornaria larvae. The larvae have rhabdomeric eyes, although adults lack eyes.

Sea angels are now recognised as the living representatives of the otherwise extinct group Graptolithoidea. Graptolites were colonies of benthic, sessile or pelagic animals found from the middle Cambrian to late Carboniferous (510 to 320 million years ago).

References: Beli et al. 2018; Cannon et al. 2014

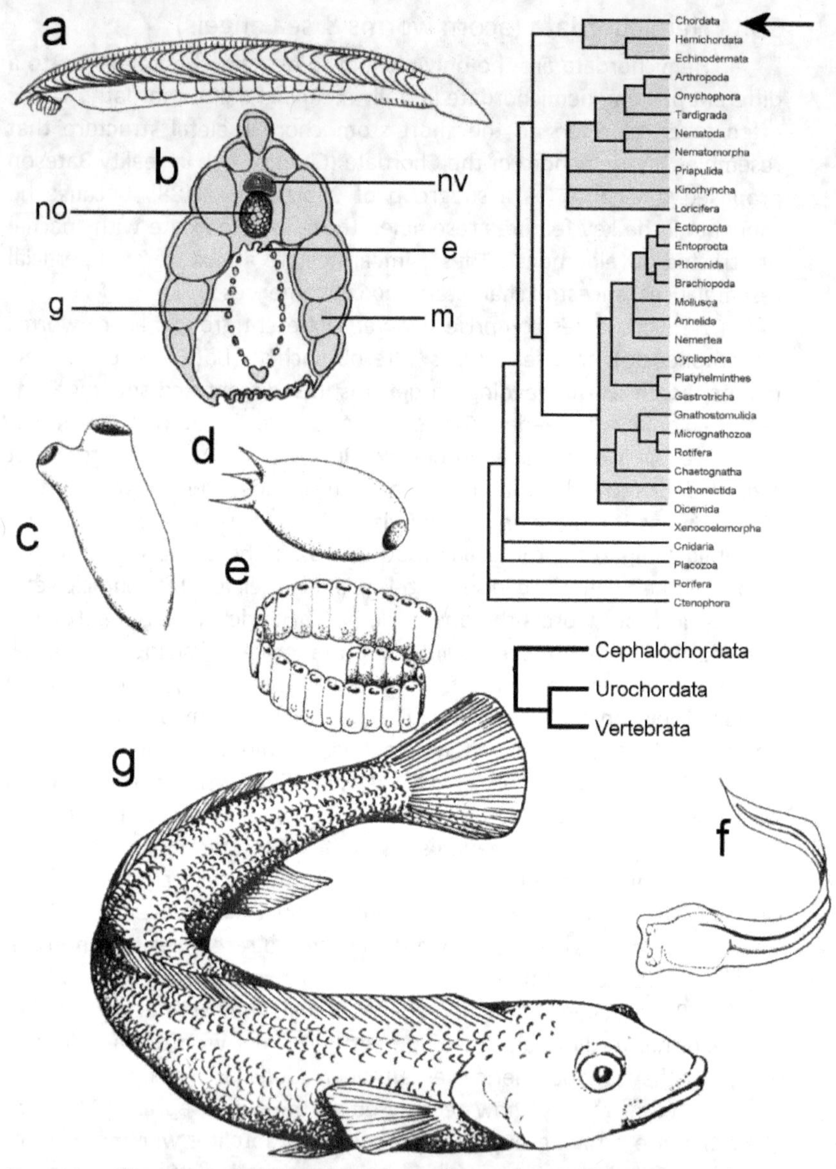

Chordata: a) Cephalochordata, b) section through a cephalochordate (e – endostyle, g – gills, m – muscle, no – notochord, nv – nerve cord), c-f) Urochordata (c: sessile, d: pelagic, e: pelagic colonial, f: larva, g) Vertebrata

34.   Chordata (lancelets, sea squirts and vertebrates)

Chordates comprise vertebrates (Vertebrata or Craniata), sea-squirts (Urochordata) and lancelets (Cephalochordata). Vertebrates are found in all environments, sea-squirts are marine filter-feeders and lancelets filter detritus in marine sediment. The diverse vertebrate life-styles result in a great diversity of species (6,000), but sea-squirts are also diverse (3,000) whereas only 33 species of lancelets have been identified.

Chordates are defined by several characters, principally a skeletal notochord, a hollow dorsal nerve cord and a branchial basket with an endostyle. The mesodermal notochord forms a stiffening rod down the back of the embryo. Vertebrates stiffen this further with cartilage and, usually, bone. In addition to its skeletal function it also signals to surrounding tissues, causing differentiation of the nerves, blood vessels, segments and muscles. Segmentation is an important chordate feature, convergently evolved in arthropods and annelids (Chapters 17, 30). Above the notochord is the epidermal hollow dorsal nerve cord which vertebrates expand into a brain. This means that the vertebrate brain is unique in being a single structure rather than a fusion of a series of ganglia. Near the head is the branchial basket and endostyle. The branchial basket constitutes the gill slit system ancestral in all deuterostomes. The endostyle is a mucus-secreting feeding structure in lancelets which accumulates iodine as a marker of developmental time. Vertebrates stop filter-feeding and use gill slits for gas-exchange; the endostyle becomes the thyroid gland, which still depends on iodine.

Chordate circulation is closed. It is basic in lancelets which lack a heart, pumping just by muscular contraction of the blood vessel walls. They also lack oxygen transporters. In sea-squirts there is a tubular valved heart and in vertebrates it is the characteristic multi-chambered heart.

Sea-squirt genomes are radically altered, with extensive shuffling, accelerated rates of genome evolution and reduction, with just 16,000 coding genes (compared to 30,000 of vertebrates). There have been many deletions of Hox genes and colinearity has been lost. In contrast vertebrates have duplicated Hox genes through several polyploidy events. Reproduction is usually sexual. Sea-squirts may be asexual or sexual hermaphrodites, and have direct development or a tadpole larva.

The fossil record goes back to the early Cambrian 519 million years ago, it is extensive for the mineralised vertebrates only.

References: Dehal et al. 2002

References

Adrianov AV & AS Maiorova. 2015. *Pycnophyes abyssorum* sp. n. (Kinorhyncha: Homalorhagida), the deepest kinorhynch described so far. Deep-Sea Res. Part II 111: 49

Aguinaldo AMA, JM Turbeville, LS Linford, et al. 1997. Evidence for a clade of nematodes, arthropods and other moulting animals. Nature 387: 489

Albertin CB, O Simakov, T Mitros, et al. 2015. The octopus genome and the evolution of cephalopod neural and morphological novelties. Nature 524: 220

Babonis LS, MB DeBiasse, WR Francis, et al. 2018. Integrating embryonic development and Evolutionary history to characterize tentacle-specific cell types in a ctenophore, Mol. Biol. Evol. 35(12): 2940

Bailly, X, I Laguerre, G Correc, et al. 2014. The chimerical and multifaceted marine acoel *Symsagittifera roscoffensis*: from photosymbiosis to brain regeneration. Front. Microbiol. 5: 498

Baker EA & A Woollard. 2010. How weird is the worm? Evolution of the developmental gene toolkit in *Caenorhabditis elegans*. J. Dev. Biol. 7(4): 19

Ball EE & DJ Miller. 2006. Phylogeny: the continuing classificatory conundrum of Chaetognaths. Current Biol. 16(15): 593

Beli E. 2018. The zoogeography of extant rhabdopleurid hemichordates (Pterobranchia: Graptolithina), with a new species from the Mediterranean Sea. Invertbr. Syst. 32(1): 100

Borgonie G, B Linage-Alvarez, AO Ojo, et al. 2015. Eukaryotic opportunists dominate the deep-subsurface biosphere in South Africa. Nat. Commun. 6: 8952

Bottino D, A Mogilner, T Roberts, et al. 2002. How nematode sperm crawl. Journ. Cell Sci. 115: 367

Bouchet P, S Bary, V Héros et al. 2016. How many species of molluscs are there in the world's oceans, and who is going to describe them? In: Héros, Strong & Bouchet (eds) Tropical Deep-Sea Benthos 29. Mém. Mus. natn. Hist. nat. 208

Boute, N, JY Exposito, N Boury-Esnault, et al. 1996. Type IV collagen in sponges, the missing link in basement membrane ubiquity. Biol. Cell. 88(1-2): 37

Brown R. 1983. Spermatophore transfer and subsequent sperm development in a homalorhagid kinorhynch. Zool. Scr. 12: 257

Cannon JT, KM Kocot, DS Waits, et al. 2014. Phylogenomic resolution of the hemichordate and echinoderm clade. Curr. Biol. 24(23): 2827

Caron J-B & B Cheung. 2019. *Amiskwia* is a large Cambrian gnathiferan with complex gnathostomulid-like jaws. Comm. Biol. 2: 164

Chen C, K Linse, JT Copley et al. 2015. The 'scaly-foot gastropod': a new genus and species of hydrothermal vent-endemic gastropod. J. Moll. Stud. 81(3): 322

Danovaro R, C Gambi, A Dell'Anno, et al. 2016. The challenge of proving the existence of metazoan life in permanently anoxic deep-sea sediments. BMC Biol. 14(43)

Decraemer W, A Coomans & J Baldwin. 2014. Chapter 1: Morphology of Nematoda. In: Schmidt-Rhaesa (ed.), Handbook of Zoology. De Gruyter, Berlin

Dehal P, Y Satou, RK Campbell, et al. 2002. The draft genome of *Ciona intestinalis*: insights into chordate and vertebrate origins. Science 298(5601): 2157

Dilly PN. 2014. *Cephalodiscus* reproductive biology (Pterobranchia, Hemichordata). Acta Zool. 95(1)

von Döhren J & T Bartolomaeus. 2007. Ultrastructure and development of the rhabdomeric eyes in *Lineus viridis* (Heteronemertea, Nemertea). Zoology (Jena) 110: 430

Edgecombe GD & DA Legg. 2014. Origins and early evolution of arthropods. Front. Palaeo. 57(3): 457

Eitel M, H-J Osigus, R DeSalle, et al. 2013. Global Diversity of the Placozoa. PLoS ONE 8(4): e57131

Fröbius AC & P Funch. 2017. Rotiferan Hox genes give new insights into the evolution of metazoan bodyplans. Nat. Commun. 8(9)

Funch P & RM Kristensen. 1995. Cycliophora is a new phylum with affinities to Entoprocta and Ectoprocta. Nature 378(6558): 711

Funch P & RA Cardoso Neves. 2018. Cycliophora. In: Schmidt-Rhaesa (ed.), Handbook of Zoology: Miscellaneous Invertebrates. De Gruyter, Berlin

Furuya H & K Tsuneki. 2007. Developmental patterns of the hermaphroditic gonad in dicyemid mesozoans (Phylum Dicyemida). Invertebr. Biol. 126(4): 295

Garm A & D-E Nilsson. 2014. Visual navigation in starfish: first evidence. for the use of vision and eyes in starfish. Proc. R. Soc. B 281: 20133011

Garwood RJ, GD Edgecombe, S Charbonnier, et al. 2016. Carboniferous Onychophora from Montceau-les-Mines, France, and onychophoran terrestrialization. Invertebr. Biol. 135(3): 179

Giribert G & GD Edgecombe. 2020. The Invertebrate Tree of Life. Princeton University Press, Princeton

Giribet G, MV Sørensen, P Funch, et al. 2004. Investigation into the phylogenetic position of Micrognathozoa using four molecular loci. Cladistics 20(1)

Giribet G, G Hormiga & GD Edgecombe. 2016. The meaning of categorical ranks in evolutionary biology. Org. Div. Evol. 16: 427

Gittenberger A & C Schipper 2008; Long live Linnaeus, *Lineus longissimus* (Gunnerus, 1770) (Vermes: Nemertea: Anopla: Heteronemertea: Lineidae), the longest animal worldwide and its relatives occurring in The Netherlands. Zool. Med. Leiden 82(7): 59

Gross V & G Mayer. 2015. Neural development in the tardigrade *Hypsibius dujardini* based on anti-acetylated α-tubulin immunolabeling. Evodevo. 6: 12

Halberg KA, A Jørgensen & N Møbjerg. 2013. Desiccation tolerance in the tardigrade *Richtersius coronifer* relies on muscle mediated structural reorganization. PLoS ONE 8(12): e85091

Han J, S Morris, Q Ou et al. 2017. Meiofaunal deuterostomes from the basal Cambrian of Shaanxi (China). Nature 542: 228

Hanelt B, F Thomas & A Schmidt-Rhaesa. 2005. Biology of the phylum Nematomorpha. Adv. Parasitol. 59: 244

Harris JE & HD Crofton. 1957. Structure and function in the nematodes: internal pressure and cuticular structure in Ascaris. J. Exp. Biol. 34: 116

Hashimoto T, D Horikawa, Y Saito et al. 2016. Extremotolerant tardigrade genome and improved radiotolerance of human cultured cells by tardigrade-unique protein. Nat. Commun. 7: 12808

Hermosilla C, LM Silva, R Prieto, et al. 2015. Endo- and ectoparasites of large whales (Cetartiodactyla: Balaenopteridae, Physeteridae): Overcoming difficulties in obtaining appropriate samples by non- and minimally-invasive methods. Int. J. Parasitol. Parasites Wildl. 4(3): 414

Hochberg FG & R Hofrichter. 2005. Dicyemida (Rhombozoa, "Mesozoa"). In: Hofrichter (ed.), El mar Mediterráneo. Fauna, flora, ecología. II/1: Guía sistemática y de identificación. Procariota, protistas, hongos, algas, animales (hasta Nemertea). Ediciones Omega, Barcelona

Hochberg R & MK Litvaitis. 2001. Phylogeny of Gastrotricha: A Morphology-Based Framework of Gastrotrich Relationships. Biol. Bull. 198(2): 299

Holland PW. 2013. Evolution of homeobox genes. Wiley Interdiscip. Rev. Dev. Biol. 2(1): 31

Hou, X, J Bergström & Y Jie 2006. Distinguishing anomalocaridids from arthropods and priapulids. Geol. Journ. 41(3–4): 259

Jager M, R Chiori, A Alié, et al. 2010. New insights on ctenophore neural anatomy: immunofluorescence study in *Pleurobrachia pileus* (Müller, 1776). J. Exp. Zool. B. Mol. Dev. Evol. 316B(3): 171

Janssen R, BJ Eriksson, NN Tait et al. 2014. Onychophoran Hox genes and the evolution of arthropod Hox gene expression. Front. Zool. 11(22)

Kristensen RM. 1983. Loricifera, a new phylum with Aschelminthes characters from the meiobenthos. Z. zool. Syst. Evolut.-forsch. 21(3): 163

Kristensen, RM. 2002. An Introduction to Loricifera, Cycliophora, and Micrognathozoa. Integrative and Compar. Biol. 42(3): 641

Kristensen RM & P Funch. 2000. Micrognathozoa: a new class with complicated jaws like those of Rotifera and Gnathostomulida. J. Morphol. 246: 1

Laumer CE, R Fernández, S Lemer, et al. 2019. Revisiting metazoan phylogeny with genomic sampling of all phyla. Proc. R. Soc. B. 286: 20190831

Lemche HA 1957. New living deep-sea mollusc of the Cambro-Devonian Class Monoplacophora. Nature 179: 413

Lindsay DJ & H Miyake. 2007. A novel benthopelagic ctenophore from 7217m depth in the Ryukyu Trench, Japan, with notes on the taxonomy of deep sea cydippids. Plankt. Benthos Res. 2(2): 98

Lozano-Fernandez, J, M dos Reis, PCJ Donoghue et al. 2017. RelTime rates collapse to a strict clock when estimating the timeline of animal diversification. Genome Biol. Evol. 9: 1320

Lu T, M Kanda, N Satoh et al. 2017. The phylogenetic position of dicyemid mesozoans offers insights into spiralian evolution. Zoological Lett. 3: 6

Luo Y-J, T Takeuchi, R Koyanagi, et al. 2015. The *Lingula* genome provides insights into brachiopod evolution and the origin of phosphate biomineralization. Nat. Commun. 6: 8301

Martín-Durán, JM, R Janssen, S Wennberg, et al. 2012. Deuterostomic development in the protostome *Priapulus caudatus*. Curr. Biol. 22: 2161

Marlétaz F, KTCA Peijnenburg, T Goto, et al. 2019. A new spiralian phylogeny places the enigmatic arrow worms among gnathiferans. Curr. Biol. 29(2): 312

Martin V. 2002. Photoreceptors of cnidarians. Can. J. Zool. 80(10): 1703

Mayer, G, PM Whitington, P Sunnucks, et al. 2010. A revision of brain composition in Onychophora (velvet worms) suggests that the tritocerebrum evolved in arthropods. BMC Evol. Biol. 10: 255

Mayer G, IS Oliveira, A Baer, et al. 2015. Capture of prey, feeding, and functional anatomy of the jaws in velvet worms (Onychophora). Integrative and Compar. Biol. 55(2): 217

Maxwell EK, JF Ryan, CE Schnitzler, et al. 2012. MicroRNAs and essential components of the microRNA processing machinery are not encoded in the genome of the ctenophore. *Mnemiopsis leidyi*. BMC Genomics 13(1): 714

Moens T, U Braeckman, S Derycke et al. 2013. Ecology of free-living marine nematodes. Handbook of Zoology: Gastrotricha, Cycloneuralia and Gnathifera, vol. 2: Nematoda. De Gruyter, Berlin

Moroz LL & AB Kohn. 2016. Independent origins of neurons and synapses: insights from ctenophores. Phil. Trans. R. Soc. Lond. B. Biol. Sci. 371(1685): 20150041

Neves RC & RM Kristensen. 2016. *Spinoloricus neuhausi* (Loricifera, Nanaloricidae), a new deep sea species from Galápagos Spreading Center. Zool. Anz. 265: 171

Nielsen C. 1971. Entoproct life-cycles and the entoproct/ectoproct relationship, Ophelia 9(2): 209

Nielsen C. 201. Animal Evolution: Interrelationships of the Living Phyla. 3rd ed. Oxford University Press, Oxford

Nishiguchi MK & R Mapes. 2008. Cephalopoda. In: Ponder & Lindberg (eds.) Phylogeny and evolution of the Mollusca. University of California Press, Berkeley

Passamaneck Y & K Halanych. 2006. Lophotrochozoan phylogeny assessed with LSU and SSU data: Evidence of lophophorate polyphyly. Mol. Phyl. Evol. 40(1): 20

Pastrana CC, MB DeBiasse & JF Ryan. 2019. Sponges pack ParaHox genes. Genome Biol. Evol. 11(4): 1250

Poinar G & R Buckley. 2006. Nematode (Nematoda: Mermithidae) and hairworm (Nematomorpha: Chordodidae) parasites in Early Cretaceous amber. Journ. Invertebr Path. 93(1): 36

Reed CG. 1991. Bryozoa. In Giese, Pearse & Pearse (eds): Reproduction of Marine Invertebrates Vol. 6. Boxwood Press, Pacific Grove

Ricci C. 2016. Bdelloid rotifers: 'sleeping beauties' and 'evolutionary scandals', but not only. Hydrobiologia 796: 277

Rouse GW. 2016. Phylum Annelida. The segmented (and some unsegmented) worms. In: Brusca, Moore & Shuster (eds.), Invertebrates, 3rd ed. Sinauer, Sunderland

Rouse GW, NG Wilson, JI Carvajal et al. 2016. New deep-sea species of *Xenoturbella* and the position of Xenacoelomorpha. Nature 530: 94

Ryan JF, K Pang, JC Mullikin, et al. 2010. The homeodomain complement of the ctenophore *Mnemiopsis leidyi* suggests that Ctenophora and Porifera diverged prior to the ParaHoxozoa. EvoDevo 1: 9

Schiemann SM, JM Martín-Durán, A Børve, et al. 2017. Clustered brachiopod Hox genes are not expressed collinearly and are associated with lophotrochozoan novelties. PNAS 114(10): E1913

Schiffer PH, HE Robertson & MJ Telford. 2018. Orthonectids are highly degenerate annelid worms. Curr. Biol. 28: 1970

Schmidt C & P Martinez Arbizu. 2015. Unexpectedly higher metazoan meiofauna abundances in the Kuril-Kamchatka Trench compared to the adjacent abyssal plains. Deep-Sea Res. Part II 111: 60

Schmidt-Rhaesa A. 1998. Muscular ultrastructure in *Nectonema munidae* and *Gordius aquaticus* (Nematomorpha). Invertebr. Biol. 117: 37

Shimizu K, Y-J Luo, N Satoh et al. 2017. Possible cooption of engrailed during brachiopod and mollusc shell development. Biol. Lett. 13: 20170254

Simakov O, T Kawashima, F Marlétaz, et al. 2015. Hemichordate genomes and deuterostome origins. Nature 527: 459

Smith FW, TC Boothby, I Giovannini, et al. 2016. The compact body plan of tardigrades evolved by the Loss of a large body region. Curr. Biol. 26(2): 224

Sørensen MV, P Funch, E Willerslev, et al. 2000. On the phylogeny of the Metazoa in light of Cycliophora and Micrognathozoa. Zool. Anz. 239: 297

Srivastava, M, E Begovic, J Chapman, et al. 2008. The *Trichoplax* genome and the nature of placozoans. Nature 454(7207): 955

Steinmetz PHR, JEM Kraus, C Larroux, et al. 2012. Independent evolution of striated muscles in cnidarians and bilaterians. Nature 487: 231

Stokes AN, PK Ducey, L Neuman-Lee, et al. 2014. Confirmation and distribution of tetrodotoxin for the first time in terrestrial invertebrates: two terrestrial flatworm species (*Bipalium adventitium* and *Bipalium kewense*). PLoS ONE 9(6): e100718

Stork NE. 2018. How many species of insects and other terrestrial arthropods are there on Earth? Ann. Rev. Entomol. 63: 31

Szaniawski H. 2002. New evidence of the proconodont origin of chaetognaths. Acta Palaeont. Pol. 47: 405

Takada N, T Goto & N Satoh. 2002. Expression pattern of the Brachyury gene in the arrow worm *Paraspadella gotoi* (Chaetognatha). Genesis 32(3): 240

Temereva EN. 2017. Innervation of the lophophore suggests that the phoronid *Phoronis ovalis* is a link between phoronids and bryozoans. Sci. Rep. 7: 14440

Tunnacliffe A, J Lapinski & B McGee. 2005. A putative LEA protein, but no trehalose, is present in anhydrobiotic bdelloid rotifers. Hydrobiologia 546 (1): 315

Warwick, RM. 2000. Are loriciferans paedomorphic (progenetic) priapulids? Vie Milieu 50: 191

Williams A. 1997. Brachiopoda: Introduction and integumentary system. In: Harrison & Woollacott (eds.) Microscopic Anatomy of Invertebrates, Vol. 13: Lophophorates, Entoprocta, and Cycliophora. Wiley-Liss, New York

Worsaae K & RM Kristensen. 2016. Phylum Micrognathozoa: The micrognathozoans. In: Brusca, Moore & Shuster (eds.), Invertebrates, 3rd ed. Sinauer, Sunderland

Yoshida Y, G Koutsovoulos, DR Laetsch, et al. 2017. Comparative genomics of the tardigrades *Hypsibius dujardini* and *Ramazzottius varieornatus*. PLoS Biol. 15(7): e2002266

Zhang H, S Xiao, Y Liu, et al. 2015. Armored kinorhynch-like scalidophoran animals from the early Cambrian. Sci. Rep. 5: 16521